Islands of Intensive Agriculture in Eastern Africa

EASTERN AFRICAN STUDIES

Revealing Prophets
Edited by DAVID M. ANDERSON & DOUGLAS H. JOHNSON

East African Expressions of Christianity
Edited by THOMAS SPEAR & ISARIA N. KIMAMBO

The Poor Are Not Us
Edited by DAVID M. ANDERSON & VIGDIS BROCH-DUE

Potent Brews
JUSTIN WILLIS

Swahili Origins
JAMES DE VERE ALLEN

Being Maasai
Edited by THOMAS SPEAR & RICHARD WALLER

Jua Kali Kenya
KENNETH KING

Control & Crisis in Colonial Kenya
BRUCE BERMAN

Unhappy Valley
Book One: State & Class
Book Two: Violence & Ethnicity
BRUCE BERMAN & JOHN LONSDALE

Mau Mau from Below
GREET KERSHAW

The Mau Mau War in Perspective
FRANK FUREDI

Squatters & the Roots of Mau Mau 1905–63
TABITHA KANOGO

Economic & Social Origins of Mau Mau 1945–53
DAVID W. THROUP

Multi-Party Politics in Kenya
DAVID W. THROUP & CHARLES HORNSBY

Empire State-Building
JOANNA LEWIS

Decolonization & Independence in Kenya 1940–93
Edited by B.A. OGOT & WILLIAM R. OCHIENG'

Eroding the Commons
DAVID ANDERSON

Penetration & Protest in Tanzania
ISARIA N. KIMAMBO

Custodians of the Land
Edited by GREGORY MADDOX, JAMES L. GIBLIN & ISARIA N. KIMAMBO

Education in the Development of Tanzania 1919–1990
LENE BUCHERT

The Second Economy in Tanzania
T.L. MALIYAMKONO & M.S.D. BAGACHWA

Ecology Control & Economic Development in East African History
HELGE KJEKSHUS

Siaya
DAVID WILLIAM COHEN & E.S. ATIENO ODHIAMBO

Uganda Now
Changing Uganda
Developing Uganda
From Chaos to Order
Religion & Politics in East Africa
Edited by HOLGER BERNT HANSEN & MICHAEL TWADDLE

Kakungulu & the Creation of Uganda 1868–1928
Michael Twaddle

Controlling Anger
SUZETTE HEALD

Kampala Women Getting By
SANDRA WALLMAN

Political Power in Pre-Colonial Buganda
RICHARD J. REID

Alice Lakwena & the Holy Spirits
HEIKE BEHREND

Slaves, Spices & Ivory in Zanzibar
ABDUL SHERIFF

Zanzibar Under Colonial Rule
Edited by ABDUL SHERIFF & ED FERGUSON

The History & Conservation of Zanzibar Stone Town
Edited by ABDUL SHERIFF

Pastimes & Politics
LAURA FAIR

Ethnicity & Conflict in the Horn of Africa
Edited by KATSUYOSHI FUKUI & JOHN MARKAKIS

Conflict, Age & Power in North East Africa
Edited by EISEI KURIMOTO & SIMON SIMONSE

Property Rights & Political Development in Ethiopia & Eritrea
SANDRA FULLERTON JOIREMAN

Revolution & Religion in Ethiopia
ØYVIND M. EIDE

Brothers at War
Tekeste Negash & KJETIL TRONVOLL

From Guerrillas to Government
DAVID POOL

Mau Mau & Nationhood
Edited by E.S. ATIENO ODHIAMBO & JOHN LONSDALE

A History of Modern Ethiopia 1855–1991
Second edition
BAHRU ZEWDE

Pioneers of Change in Ethiopia
BAHRU ZEWDE

Remapping Ethiopia
Edited by W. JAMES, D. DONHAM, E. KURIMOTO & A. TRIULZI

Southern Marches of Imperial Ethiopia
Edited by DONALD L. DONHAM & WENDY JAMES

A Modern History of the Somali
Fourth edition
I.M. LEWIS

*Dhows & the Colonial Economy in Zanzibar 1860–1970**
ERIK GILBERT

*In Search of a Nation**
Edited by GREGORY H. MADDOX, & JAMES L. GIBLIN

*A History of the Excluded**
JAMES L. GIBLIN

Islands of Intensive Agriculture in East Africa
Edited by MATS WIDGREN & JOHN E.G. SUTTON

*African Womanhood in Colonial Kenya**
TABITHA KANOGO

Leaf of Allah
EZEKIEL GEBISSA

* forthcoming

Islands of Intensive Agriculture in Eastern Africa

Past & Present

Edited by

MATS WIDGREN
Professor in Geography at Stockholm University

&

JOHN E.G. SUTTON
Former Director of the British Institute in Eastern Africa

The British Institute in Eastern Africa
& Stockholm University

in association with

E.A.E.P.
NAIROBI

Mkuki na Nyota
DAR ES SALAAM

Ohio University Press
ATHENS

James Currey
OXFORD

The British Institute in Eastern Africa
10 Carlton House Terrace
London SW1Y 5AH
& P.O. Box 30710
Nairobi

Department of Human Geography
Stockholm University,
S–106 91 Stockholm
ISSN 0585–3508
No 121 in the Meddelanden Series

in association with

James Currey Ltd
73 Botley Road
Oxford
OX2 0BS

Ohio University Press
Scott Quadrangle
Athens, Ohio 45701

Mkuki na Nyota
P.O. Box 4205
Dar es Salaam

East African Educational Publishers
Mpaka Road/Woodvale Grove
P.O. Box 45314
Nairobi

First published 2004

British Library Cataloguing in Publication Data
Islands of intensive agriculture in Eastern Africa : past & present. -
(Eastern African Studies)
1. Agricultural intensification - Africa, Eastern
2. Agriculture - Africa, Eastern - History 3. Agriculture - Social aspects
- Africa, Eastern 4. Africa, Eastern - Rural conditions
I. Widgren, Mats II. Sutton, John E. G. III. British Institute in Eastern
Africa IV. Stockholms universitet
631.5'8'09676

ISBN 0-85255-427-3 (James Currey Cloth)
0-85255-428-1 (James Currey Paper)

Library of Congress Cataloging-in-Publication Data
A catalog record for this book is available on request from the Library of Congress

ISBN 0-8214-1561-1 (Ohio University Press Cloth)
0-8214-1562-X (Ohio University Press Paper)

Typeset in 10/11pt Baskerville
by the British Institute in Eastern Africa
Printed and bound in Great Britain
by Woolnough, Irthlingborough

Contents

List of Figures & Tables

Figures

Tables

Notes on Contributors

William M. Adams, PhD, is Fellow of Downing College, Cambridge, and University Reader in the Geography of Conservation and Development. He has extensive experience of agricultural and irrigation systems in both western and eastern Africa. His publications include *Green Development: Environment and Sustainability in the Third World* (new edition 2001) *and Wasting the Rain: Rivers, People and Planning in Africa* (1992).

Lowe Börjeson, Fil.lic., is a doctoral student at the Department of Human Geography, Stockholm University. A field-oriented historical geographer with experience in Sweden, Ireland and Botswana, he is currently researching the Iraqw agrarian community in Tanzania.

Vesa-Matti Loiske, PhD, lectures in Human Geography and directs the Coastal Management Research Centre at Södertörns högskola (University College, South Stockholm). From 1991 to 1997 he worked on social issues related to land management among the Iraqw of north-central Tanzania. His publications include *The Village that Vanished: the Roots of Erosion in a Tanzanian Village* (1995).

Wilhelm Östberg, PhD, is Docent in Social Anthropology at Stockholm University and curator in the National Museum of Ethnography, Stockholm, with extensive experience in southern and eastern Africa. His books include *The Kondoa Transformation: Coming to Grips with Soil Erosion in Central Tanzania* (1986) and *Land is Coming up: the Burunge of Central Tanzania and their Environments* (1995).

John E.G. Sutton, PhD FSA, has a special interest in the African Iron Age and the history of agricultural systems. He was formerly Professor of Archaeology in the University of Ghana and, from 1983 to 1998, Director of the British Institute in Eastern Africa and editor of the journal *Azania*. His publications include *A Thousand Years of East Africa* (1990) and *Archaeological Sites of East Africa* (1999).

Elizabeth E. Watson, PhD, lecturer in Geography at Cambridge University, works on social and cultural change and rural livelihoods in Africa, and has field experience in Kenya, Ethiopia and Mozambique. Her doctoral thesis, 'Ground Truths: Land and Power in Konso, Ethiopia', was presented in Cambridge in 1998.

Mats Widgren, PhD, is Professor in Geography at Stockholm University. His principal research has been on the history of agricultural landscapes from the Iron Age to the present both in Sweden and in southern and eastern Africa. His publications include *Settlement and Farming Systems in the Early Iron Age* (1983).

Preface

The understanding of African agricultural practices has advanced in recent years through geographical, ethnographic and archaeological research, concentrating on the actual fields, farming strategies and cultivation techniques of both existing and former communities. Certain of these communities, situated along the line of the East African Rift Valley and in the highlands to either side, have developed highly specialized agricultural regimes with labour-intensive devices, such as furrow irrigation, hillside terracing to conserve soil, and stall-feeding of cattle to supply manure, all these combined with specified crop rotations and maintained by complex communal rules of land use and labour mobilization. Such agricultural systems, in which repertoires of special techniques may be integrated, often occur in remote hill districts or may be surrounded by dry and sparsely populated terrain. Each example is different in detail, the peculiarities, social as much as strictly agricultural, usually being geographically circumscribed and adapted, albeit in unpredictable ways, to the local conditions (of soil, rainfall, water availability, etc). Although it is clear that all these individual communities have considerable historical depth – with moreover the archaeological remains of the irrigated field system of Engaruka dating back some five or six centuries – there is little suggestion of specific connections or migrations over long distances; and the old idea of an ancient and widely diffused 'Rift Valley civilization' has long been discredited. Rather, these systems, their similar features as well as their contrasts, seem to have evolved essentially in their own locations. Some began with the advantage of plentiful rainfall; more often that is inadequate or unreliable so that water-conservation devices or irrigation have been developed. Commonly they exhibit an air of self-sufficiency and relative isolation, in a social as well as a purely physical sense, although this may in fact mask the degree of exchange and interaction, if not population movement, occurring with surrounding pastoralists or with cultivators in neighbouring districts.

It is this phenomenon that we recognize as *islands* of intensive agriculture. Attention was drawn to this subject (and to the intensification debate) in a special volume (No. 24) of the journal *Azania* for 1989, *The History of African Agricultural Technology and Field Systems*. Since then new directions of cross-disciplinary fieldwork and enquiry have been pursued by, among others, teams from the Department of Human Geography of Stockholm University and the British Institute in Eastern Africa, with valuable contributions by colleagues in several other institutions.

The present volume documents some of the results of this recent research and of the collaboration between these institutions. This was facilitated by grants from the Swedish International Development Authority (SIDA) and the Swedish Council for Planning and Coordination of Research, which enabled the Stockholm team to expand their fieldwork in the late 1990s and to coordinate with local institutions, in both Tanzania and Kenya. Alongside this, a seminar was arranged in Nairobi at the British Institute in November 1998, preliminary to a panel organized for the conference held at Oxford the following July, 'African environments, past and present'. We are grateful to the organizers of that conference, Professor William Beinart and Dr JoAnn McGregor, for providing the opportunity for this specialist panel and agreeing to its proceedings being published separately, and to our Cambridge colleague, Bill Adams, for acting as discussant and, subsequently, preparing a concluding chapter to this collection. We are also appreciative of the continuing support for the project from the British Institute in Eastern Africa and from its current director, Dr Paul Lane; and equally from the Leverhulme Trust which provided an emeritus fellowship to one of us, held at BIEA and the Pitt Rivers Museum in Oxford, for pursuing this work in the years 1998–2000, after handing over the direction of BIEA.

Earlier versions of most of the papers in this collection were prepared for the Oxford conference and bound in 1999 as Working Paper No. 43 of the Environment and Development Studies Unit of Stockholm University. The authors and editors thank all colleagues who, during that conference and subsequently, offered constructive comments, both on the individual studies and on the overall theme.

J.E.G.S./M.W.

One

Towards a Historical Geography of Intensive Farming in Eastern Africa[1]

MATS WIDGREN

In the literature on eastern Africa, instances of intensive agriculture[2] are documented in a number of localities. Most of these – between the Great Lakes, in the Kenya Highlands, and on the slopes of Kilimanjaro and other mountains of northeastern Tanzania – are well watered, and many of them have a long tradition of banana cultivation. But also in less favoured environments, with medium or low rainfall, farmers have over the generations intensified farming and created permanent garden-like landscapes, by such devices as construction of irrigation furrows, levelling the fields, sometimes with stone terracing, or applying mulch or the manure from their livestock. These labour-intensive and productive farming systems are often surrounded by areas of sparse population relying on extensive cultivation or pastoralism; hence the metaphor 'islands of intensification' (Maggs 1995).[3]

Farming communities employing such intensive agricultural practices were noticed by some of the first European travellers in East Africa in the late nineteenth century. Subsequently, during the colonial period, some of these farming systems were described in detail by agricultural officers who were impressed by the complexity and versatility of their techniques for successfully countering poor soils and uncertain rainfall. Towards the end of the twentieth century they began attracting renewed interest in development circles, as locally developed answers to what was becoming increasingly recognized as the African food crisis.

Over the years they have been the subject of a variety of academic approaches. Geographers with demographic interests have highlighted the uneven distribution of population in eastern Africa and identified population density islands. Environmental historians have looked at the same phenomenon, albeit through slightly different glasses, and debated the importance of both extensive and intensive farming practices in precolonial times and of the individual communities that have pioneered innovations or maintained intensive techniques. Archaeologists, agronomists and agricultural geographers have focused more on the physical investments of intensive farming, in particular irrigation works and ridging and levelling of fields, especially with stone-built terracing. Such investments, called 'landesque capital' (see Brookfield 1984), have modified the land surface to the extent that the settlements and their fields which eventually suffered abandonment may remain visible in an archaeological sense centuries later.

The first part of this chapter reviews these different strands of research. A definition of the object of study will be reached step by step by using the research results emanating from these approaches. Finally the different explanatory frameworks that have been used in the study of such islands, their location and their development, will be explored.

Population islands

In 1936 Clement Gillman published the first detailed population map of Tanganyika (see Figure 4.11 of Börjeson's paper). In his comments he described some of the densely settled areas in the dry lowlands as cultivation steppes, seeing them as 'veritable oases within either very sparsely inhabited or entirely uninhabited land. Nearly everywhere the change from high population density to zero is a sudden jump rather than a gradient.' (Gillman 1936: 361)

In searching for the underlying causes of this uneven population distribution, Gillman carried environmental determinism to its extreme. Not only did he regard the large populations in the highlands to be due to environmental characteristics, but he further suggested simple explanations for the markedly varied population densities within lowland areas with low rainfall: 'Why this astonishing and sudden change under uniform climate with its sufficient though unreliable seasonal rainfall of 600 to 900 millimeters? The answer is simple. Presence or absence of a permanent domestic water supply in the form of comparatively shallow ground water within reach of primitive tools and skill dictates the pattern of population' (Gillman 1936: 361).

Gillman's line of argument was soon criticized by Ernst Nowack, a geographer with field experience in Tanganyika. He opposed Gillman's underlying assumption that population in Tanganyika had by that time filled all possible areas for settlement. Nowack substantiated his opinion with specific cases drawn from Gillman's population map, and supported his argument with a lively description of his own experience of water availability in different parts of Tanganyika. In certain areas where water was available in abundance, even during the dry season, he found only scattered settlements. In other areas, where 'water had to be collected drop by drop from deep rock crevices, laboriously raised from a great depth and then carried far on steep mountain paths', he found village after village and a population with (what he regarded as) advanced culture.[4]

Nowack furthermore claimed that the present population pattern was dynamic and reflected fluctuations and population movements. Dense populations could, according to Nowack, be explained by historical factors (migrations, war, refuge, trade routes and influence of powerful chiefs), while empty and thinly settled lands were often those which had suffered depopulation and had not yet been recolonized. Nowack's perceptive view of the strongly dynamic character of population patterns in Africa is in close agreement with that of later researchers. His view finds support in McCann's recent synthesis of the environmental history of Africa:

> The issue for human impact on an African historical landscape is the number of people at a place at a certain time – population conjuncture – and not a simple growth rate. After all, people move. They migrate to or leave an urban area and congregate on productive land, or are coerced to do so, at a rate much higher and more significant than an overall population growth rate. The historian's task then is to detect evidence of movement and density of human population. The effect of population is therefore situational and fluid and not amenable to generalization. (McCann 1999: 20)

This highly irregular pattern of population distribution was confirmed by Trewartha's and Zelinsky's population map of most of tropical Africa published in 1954, showing that Gillman's observations from Tanganyika represented a trait general to the whole region. There was a great variety in population patterns, not only between dry and high-rainfall areas, but also within zones of similar rainfall pattern. In particular within the semi-arid zone distinct discontinuities could be documented. In eastern Africa 'the element of contrast dominates the population map' (Trewartha and Zelinsky 1954:143). The distribution was described as having an 'extreme irregularity' and 'abrupt alternation of unusually high densities with virtually unpopulated land' (op.cit.: 143). As they pointed out, the diversity in physical geography of eastern Africa, in particular the great variability in altitude, has affected this population pattern. Trewartha and Zelinsky refrained, however, in their analysis from any simple environmental determinism, and referred to Gourou's findings (see pp. 13-14) in different parts of Africa to show the role of culture in determining the population pattern.

The arguments advanced in the various geographers' reflections on population maps thus place the problem of the 'islands' centrally in a classic debate within geography, that between environmental determinism and possibilism. For some readers it may seem an obsolete discussion, but the fundamental questions continue to be relevant for current scholarship on environment, population and development and, not least, for the policies advanced by donor agencies, NGOs and intergovernmental organizations. Blaut (1993) has exposed the general aspects of the environmental determinist heritage. He argues that the environmental determinism of the late nineteenth century still has, through the works of Marx and Weber, a profound influence on modern social theory.

But in a more concrete fashion the issue has been raised in relation to problems of direct relevance to the issue of this book. Morgan, in a comparative study of Tamilnad in southern India and eastern Tanzania, focused on the problem of why 'lands in different parts of the world, but with closely comparable environments, may be used very differently by their inhabitants' (Morgan 1988: 69). He asked why on the one hand Tamilnad, 'an intensely domesticated landscape' based on irrigation, and the other hand eastern Tanzania, with a population density of less than a twentieth of that of Tamilnad, have developed in such different directions despite the similar or closely comparable physical environments. His argument is that history, contacts, trade relations, etc are more important than differences in climates and soils. Morgan's possibilist arguments have recently been countered by Gunnel, who, without taking an environmental-determinist standpoint, would nevertheless put more emphasis on the actual differences in the natural environments when discussing the potential for development in Africa generally as compared with Asia (Gunnel 1997). Morgan's comparison of Tamilnad and Tanzania raises fundamental questions relevant to the subject of this volume.

If we accept Morgan's thesis that a Tamilnad type of development is – or was – possible in large parts of the African continent, then why did it not occur? Why did the examples of small-scale irrigation and intensive agriculture studied in this volume not develop into larger systems like those in Tamilnad? What were the cultural, social and environmental constraints?

Grain-based intensive agriculture

Knowledge of the role that intensive farming played in African history still seems to be confined to a rather small group of environmental historians and to researchers within the field of indigenous knowledge. All too frequently in studies of agriculture in Africa, even very recent ones, the assumption persists that, before colonization, it remained undeveloped if not stagnant. This notion of a very short agricultural history seems a particularly prevalent premise underlying much current research on soil fertility in the continent.[5] In this area of study an awareness of the historical depth of African agriculture is seldom demonstrated.[6] Scoones and Toulmin, summarizing recent research on soil fertility on the continent, put the finger on this, calling it a 'fundamental puzzle': 'If things are so bad at an aggregate level (and apparently have been for some time – agriculture has after all been practised in many areas of Africa for centuries, without external input of fertilizer), how is it that farming persists at all?' (Scoones and Toulmin 1998: 259)

A crucial point in many misunderstandings of precolonial African agriculture is the loose way in which the term 'shifting cultivation' is still used. In some general studies one finds all precolonial cultivation simply branded as 'shifting', despite the many calls for a more specific terminology to describe the broad variety of extensive farming systems known in the tropics. (For terminology see, for example, Manshard 1968: 83ff., Ruthenberg 1971.) It is now well known that intensive agricultural practices based on permanent fields were widely spread in the region well before colonialism (Kjekshus 1996: 29–50, Koponen 1988a: 220–41, McCann 1995). Intercropping, mulching, ridging, manuring and irrigation have a long history and were not restricted to the high-rainfall areas with banana cultivation.

Using Gillman's map discussed above, Kjekshus showed a correlation for Tanzania between population density and agricultural innovations (1996 [1977]: 46). In 'islands' of high population density, farming was also more intensive. Kjekshus listed a number of agricultural innovations 'that had taken most of the Tanganyikan husbandmen far beyond the adaptive stage of shifting cultivation' (1996: 47). He also went so far as to claim that shifting cultivation was 'probably operated by only a small minority of peasants in precolonial Tanganyika' (1996: 27). He did not, however, probe beyond the simple dichotomy between shifting and permanent cultivation, and so, like many before him, he failed to appreciate the immense variety of extensive farming. His classification of farming systems therefore did not fully serve his purpose.

Since Kjekshus' original publication in 1977, the picture of agriculture in Tanganyika in the nineteenth century has been somewhat modified by Koponen (1988a: 220–41). By distinguishing two 'ideal' types of agriculture, he offered a more balanced overview of the different farming practices in Tanganyika before colonization: 'The ideal types were, first, grain cultivation with a long fallow period, and second, banana growing. In reality, grain cultivation systems observed fallow periods of varying lengths, and grains and banana could and did coexist in different combinations with each other and with other crops' (p. 221).

Of course, the intensity of farming was (and is) at a broad level closely related to precipitation patterns and to topography. Especially prominent in this respect are the concentrations of population around Lake Victoria, in the Kenya Highlands and on the slopes of the mountains of northeastern Tanzania. In the last region, especially Kilimanjaro and the North Pare Hills, banana cultivation clearly has a long history (estimated by De Langhe et al. 1995, from the diversity of cultivars, to be over 1,000 years). Several of the same areas saw the successful establishment of coffee as a commercial crop in the twentieth century (Bart et al. 1998).

The agricultural history of the highlands in north-eastern Tanzania has been the subject of several studies, which provide important themes and comparative detail for understanding intensive agriculture in other parts of eastern Africa. Irrigated agriculture has a long history here, combined in places with stalling of cattle and the use of manure on the fields (Kimambo 1996). It is documented on the slopes of Meru (Spear 1997: 125–7) and Kilimanjaro (Pike 1965, Masao 1974, Grove 1993) and in the mountains of North Pare (Sheridan 2001, 2002), South Pare (Hakansson 1995, 1998) and Usambara (Feierman 1990: 65, Conte 1994).

Agriculture in these highlands of Tanzania was characterized by bananas as the staple, but complemented by grains and other annual crops grown in fields at altitudes both above and below the banana gardens. Moreover, while irrigation was not a prerequisite for farming in these well-watered areas, it assisted, together with manuring, the growth of bananas in the drier months, and also facilitated two or three harvests of annual crops (Kimambo 1996: 72).

In the 1930s the settled slopes of Mount Meru and Kilmanjaro had a population density of more than 150 persons per square km (Gillman 1936). But, as Gillman pointed out, even away from these well-watered highlands, population density tended to cluster into smaller or bigger 'islands', many of which were based not on bananas but on grain cultivation. It is not always possible to draw a clear distinction between areas in which dense population is associated with high rainfall and fertile soils and those where the success of agriculture depends entirely on labour investments in land. On some of the lower slopes of the mountains, for example, grain cultivation dependent on irrigation existed already in the nineteenth century (for South Pare see Hakansson 1995: 304).

Koponen's distinction between banana-based and grain-based economies offers, however, at a general level, a workable definition by which islands of intensive agriculture can be clearly distinguished from the landmasses and peninsulas of high rainfall and high population density.

The areas discussed in this volume all fall into this intermediate group between the banana cultivating highlands and the extensive grain cultivation in savanna environments. In this group, sorghum, finger-millet and, more recently, maize constitute the staples, though bananas may also be cultivated, depending on irrigation. Furthermore, the high productivity relies to a large extent on the input of labour. It is on these grain-based, intensively farmed and populated areas that we focus in this volume.

Soil and water conservation

Within these areas dependent, contrasting farming practices were – and are – at hand. For not all areas of current high population densities have intensified their agriculture through soil-conservation methods, as pointed out by Hakansson (1989: 14), citing examples from Kenya and Tanzania. Those where intensified agriculture has caused soil erosion, owing to lack of soil and water conservation measures, include Kondoa and central Ugogo, both in Tanzania. These clearly show up as population islands in Gillman's map, but figured in the second part of the twentieth century not as successful and sustainable islands of intensive agriculture, but as areas with soil-erosion problems. In 1973 a soil conservation project was established in the Kondoa area and six years later 90,000 livestock were removed (Mung'ong'o 1995, 1999). Similarly in Mvumi Division in Ugogo livestock were evicted in 1986 (Holtland 1994).

But certain other areas with dense populations of grain farmers exhibit the most clearcut examples, both existing and historical, of intensive and sustainable farming practices, including terracing, irrigation, manuring and mulching. There are notable examples in Tanzania at Ukara Island, Matengo and the Iraqw area in the Mbulu Highlands; Marakwet and the Taita Hills in Kenya and Konso in Ethiopia are further cases (see Figure 1.1 and references in caption; also the map and table in Ludwig 1968: 92–5). In these areas farming is often a year-round undertaking and a lot of labour is invested in soil and water conservation (for Ukara, see Ludwig 1968; for Iraqw see Chapter 4).[7]

The grain-based islands of high population density, as defined above, thus have a number of possible trajectories. It remains a challenge to understand the reasons for the variety of developments. As with the problem of population density, there are no simple geographical explanations of the differences.

LANDESQUE CAPITAL

The high productivity of land and the robust nature of agricultural production in these areas depend on the application of different combinations of farming practices: manuring, composting, terracing, cut-off drains, a variety of irrigation techniques and a diverse crop repertoire. Furthermore, in several instances, there is evidence of careful management of trees and woodlands and of a high level of agro-diversity.[8] Irrigation and soil conservation are connected with investments in land, what has been termed landesque capital, defined as activities on land (and vegetation) that reach beyond the immediate needs of the coming cropping season (Blaikie and Brookfield 1987: 9). The durable structures resulting from these capital investments are also of crucial importance for the archaeological identification of past agricultural societies of that kind.

The word 'capital' might convey a misleading impression of, in Doolittle's words, 'inputs applied over short, discrete periods of time, and often to include planning, engineering expertise, and socially coordinated effort' (Doolittle 1984: 124). But, as he has shown in his study of locally developed irrigation systems in Sonora in Mexico, investments in land may result gradually from repeated small capital improvements. In this Mexican case there was no communal, large-scale, short-term construction and 'only minimal formal or long range planning is involved' (p. 134). Instead of being the result of major planning phases and investments, these landscapes are, in Doolittle's argument, the accumulated results of labour over a long time period; 'construction, cultivation and maintenance are inextricably mixed activities' (p. 135).

The case studies of continuing intensive farming described in this volume exhibit different patterns of interplay between incremental and more systemic or structural changes. For instance, in Marakwet in Kenya, as illustrated in Chapter 2, the investments in new irrigation furrows and in major repair

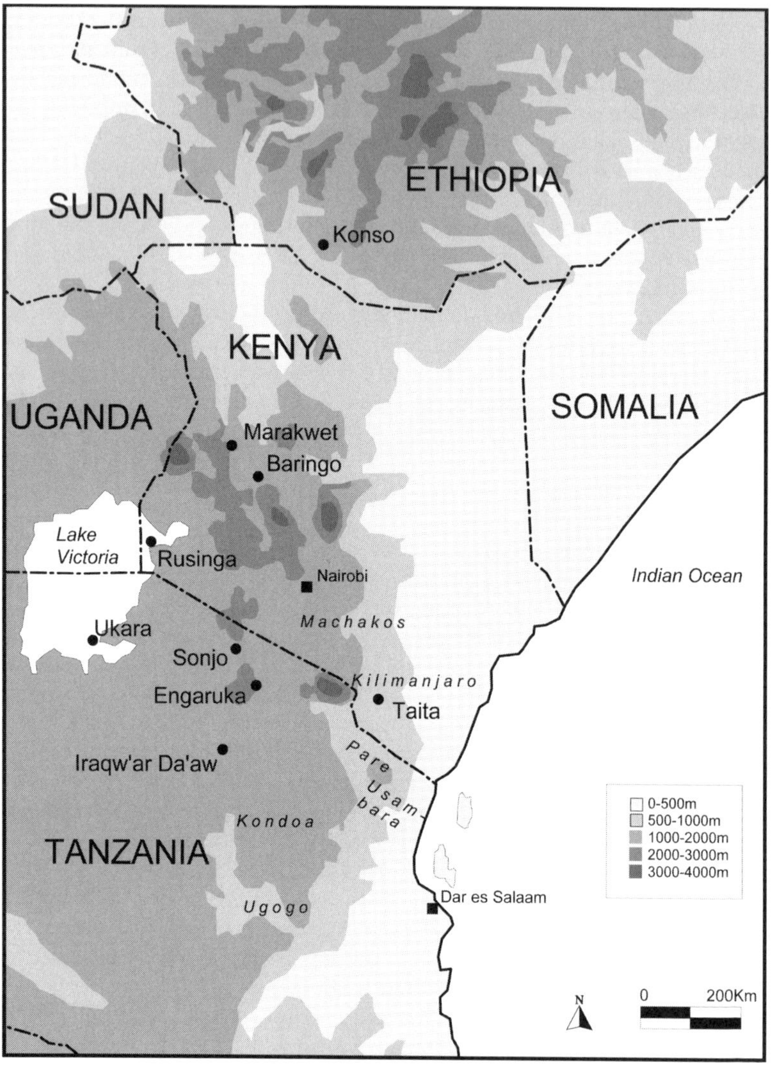

Figure 1.1 *Grain-based islands of intensive agriculture in eastern Africa, 'living' and abandoned. Full references to the various areas named can be found in the following works: Konso – Watson, Chapter 3 in this volume; Marakwet – Östberg, Chapter 2 in this volume; Baringo – Anderson 1989; Rusinga – Conelly 1994; Machakos – Tiffen et al. 1994; Ukara – Ludwig 1968; Sonjo – Adams et al. 1994; Engaruka – Sutton, Chapter 6 in this volume; Taita – Fleuret 1985 and Iraqw´ar Da/aw – Börjeson, Chapter 4 in this volume. Other areas discussed in this book are shown in italics on the map*

works are indeed capital investments, requiring a large number of people and major decisions. But contrary to many assumptions about large systems, this does not require a hierarchical organization. Konso is an illustration of another aspect of labour organization. Here major repair works and laborious investments in building new terraces are handled by the household or the communal work groups which that household is able to assemble. These work groups of various kinds, as described by Watson (Chapter 3), provide means of constructing or maintaining terraces over a few days, work which would take weeks or months if left to the labour of a single household. At the other extreme is the Iraqw case (see Chapter 4) where the cultivation and construction of terraces are 'inextricably mixed activities'. Here individual farmers make investments in the fields, in the form of cuts, terraces and drainage, as part of their daily and yearly work, and the landesque capital is thus the result of generations of tilling and sculpting the landscape.

The flexibility, the low cost and the incremental nature of this kind of investment are emphasized by Scoones et al. (1996) in an overview of indigenous soil and water conservation in Africa. The methods applied tend to spread out the labour requirements for construction and maintenance, in contrast with 'most introduced techniques which require major investments of labour in construction, often during a single period' (Scoones et al. 1996: 13). These measures, especially those aimed at water harvesting and irrigation, may also, in their view, offer comparatively fast returns to investments (op.cit.: 18).

ANTIQUITY OF IRRIGATION AND SOIL CONSERVATION

Such locally developed soil- and water-conservation structures were documented copiously during the twentieth century (see, for example, Allan 1965, Ludwig 1968, Scoones et al. 1996). However, in tracing the history of these farming practices and their resulting structures, the data are often rather imprecise. There is always a risk of the argument floating in a vague ethnographic present rather than deriving from an examination of the actual history of physical structures or of farming practices. In many cases early documentation from travellers shows that irrigation structures, terraces, labour-intensive farming and manuring were present in the late nineteenth century, and it is clear that they had already existed for quite some time.[9]

What is, however, seldom documented in the early accounts is the extent of irrigation systems and the degree of labour and cropping intensity. As described in Chapters 2 and 4, there are strong indications of the expansion of irrigation systems and of increased labour intensity during the twentieth century, in comparison with the landscapes seen by the early European travellers. Therefore the method, which some writers have followed, of simply correlating recent documentation of a 'system' with travellers' accounts of the same 'system' dating a century or more earlier, is not necessarily a reliable procedure in reconstructing landscape history. Agricultural landscapes change as do farming systems, and caution should be exercised in equating farming systems with landscapes. This is one of the reasons why the term 'farming system' is avoided in some of the papers in this volume. Physical structures can – at least in principle – be dated, as can also, albeit more vaguely, farming practices and the use of different tools and crops.

More detailed discussion of the dating of irrigation is possible for the eastern mountains in Tanzania. In South Pare Baumann's observation of old, abandoned furrows indicates that irrigation was practised well before his visit in 1890 (Hakansson 1995: 303). In the Kilimanjaro area Masao (1974) dates some of the furrows to the early nineteenth century, based on oral history, but he argues that the irrigation of Kilimanjaro may well go back to the first settlement of Chagga in the area, which he estimates at three or four centuries ago. A more detailed documentation by Tagseth (2000: 92) of the history of a set of furrows on Kilimanjaro produces an estimated origin for that particular example in the 1840s.[10]

In Usambara, oral traditions place the construction of some of the furrows to a period before the founding of the Kilindi kingdom in the eighteenth century (Feierman 1990: 65). And more generally for this region Kimambo (1996: 73) has argued that 'from Usambara to Meru, each irrigation furrow was identified with a lineage which had first planned or constructed it', thus implying that the practice of irrigation predates the period of political centralization.

The clearest direct evidence of the antiquity of intensive farming practices and of soil and water conservation in eastern and southern Africa derives from abandoned field systems (for an overview see Sutton 1984, 1985, Grove and Sutton 1989). At Engaruka, at the foot of the Rift escarpment in northern Tanzania, the extensive archaeological remains testify an extensive system of canals and irrigated fields covering some 2,000ha. Settlement remains and stone enclosures integrated in the field system indicate that cattle were stall-fed and manure was collected. The Engaruka settlement dates from about the fifteenth century to the seventeenth or possibly early eighteenth century (see Chapter 6).

Looking beyond the strict limits of eastern Africa, there are other examples of the antiquity of such intensive farming practices. At Nyanga in the eastern highlands of Zimbabwe a vast complex of ancient fields and settlements remains visible, covering over 19,000ha. The indications of intensive farming practices there consist of terraces, cultivation ridges, furrows for irrigation and water control, and of pit-structures most likely designed for cattle and the collection of dung for manure (Soper 2000, 2002). Soper's recent investigations confirm Summers' (1958) earlier findings on the dating of these structures. The main features belong to between roughly the seventeenth and early nineteenth centuries (Soper 2000: 232, 1996: 24).

The most southerly known example of abandoned precolonial intensive agriculture is in an area called Marateng around Lydenburg in the northern Drakensberg in South Africa. It is mentioned briefly by Summers (1958: 250) and has more recently been examined by Maggs (1995 and pers. comm.). This consists of a huge complex of terraces, field divisions and occasional water furrows. The arrangement of droveways (cattle paths) leading through the fields into stone enclosures (kraals) is a clear indication that it was an integrated system with cattle dung being used to manure the fields. Although its dating is not precisely established, it seems that this extensive terracing may have continued in use into the nineteenth century and that its abandonment may even be directly connected with the Boer settlement of the area in the late nineteenth century.

Considered together, these archaeological remains provide evidence that manuring, terracing and irrigation have been known and practised in certain parts of sub-Saharan Africa for several centuries. Their abandonment raises of course questions about the dynamics of intensive agriculture, a question to be taken up below.

The research problem

The discussion above has served to demarcate the phenomena we are studying. We are looking at sustainable and intensive agriculture in situations not affected by a strong market presence and not simply responding to recent population increase. All the areas covered in this volume have (or had) a population density above that of those surrounding them; moreover they rely (or relied) on grain as their staple, while landesque capital play (or played) an important role in ensuring stable and persistent production.

The argument that these islands represent sustainable farming and sustainable societies is based mainly on historical reasoning, rather than on any ecological modelling or measurements of soil fertility. This claim for sustainability may appear to be contradicted by the experience of Engaruka, and the two southern African examples noted, which did not persist till the present. Therefore some qualification or explanation is required.

To start with the examples of living islands of intensive farming – Iraqw, Marakwet and Konso – these have all been able to support a fairly large population for a long period. To judge from their historical experience, they have been less fragile than surrounding areas, in the face of both drought and human disturbances. Some of the cases illustrated in this volume may be poorer than their more expansive peripheries, but they still – through traditional networks of exchange – play an important role in the regional food security system.

There is ample evidence from the literature to show that these islands have been acting as safety valves for neighbouring areas in times of famine. In Marakwet, for instance, a colonial officer commenting upon the poor food situation in 1940 reported that they 'have been saved by their furrows' (quoted in Adams 1996: 159). In Mbulu District in Tanzania, colonial officers in the 1920s and 1930s were well aware of the role of Iraqw'ar Da/aw (Mama Issara) in relation to the surrounding areas. They supplied grain to the Mbugwe in the east and also to drought-stricken farmers to the west

(Snyder 1996: 323). A British district officer in 1922, admiring the agricultural skills of the Iraqw, noted that: 'in good years there is a large surplus of food. Thousands of lives have been saved in neighbouring districts in the past by Mbulu skill and industry' (Mbulu District Book, 'Tribal History and Legends'). Hakansson has highlighted similar relationships between areas of intensive farming and surrounding pastoralist or agro-pastoralists among the Gusii in Kenya (Hakansson 1994: 270) and Pare in Tanzania (Hakansson 1995).[11]

The question whether the abandoned farming communities noted above – Engaruka, Nyanga and Marateng – were sustainable raises, of course, fundamental questions about the time scales of social sustainability. During their lifetimes these areas probably played roles similar to those of the surviving communities studied here. The archaeological findings furthermore indicate that these were not ephemeral experiments, but that the farming communities used the same fields and the same landesque capital for several hundred years. The evidence on the ground demonstrates that they had ways of coping with water management and fertility maintenance. Brookfield, discussing such abandoned farming landscapes in a global perspective, has argued against environmental degradation and overpopulation as the prime factor in the abandonment, asserting that 'the capital-intensive systems were almost all highly conservationist, and it was their breakdown and abandonment that was more likely to yield damage to the land' (Brookfield, 1986: 180).

Explanations of intensification

As discussed above, there are no simple geographical factors which can satisfactorily explain the location and the development of these islands of intensive farming. The distribution of intensive agriculture in eastern Africa is thus not a direct reflection of natural conditions, but of a complex interaction of ecological, social and historical factors.

ETHNIC CONNECTIONS

In attempts to understand the historical background to the different areas of intensive farming in the region, the linguistic and ethnic connections between the different groups practising irrigation or terracing have been discussed in the literature. The case of Engaruka, with an abandoned agricultural community of unknown ethnic affiliation, has often been taken as a starting point.[12] Sutton attempted an overview of the different candidates for a close relationship to the abandoned Engaruka complex in 1978 (pp. 67–8); and in 1986 he more specifically discussed the relations between Engaruka and the Bantu-speaking Sonjo, presently practising a comparable type of irrigation agriculture. Quoting work by Ehret, he was inclined to see a connection between the Sonjo and the Cushitic-speaking Iraqw, which opened up a possibility for seeing a linguistic connection between the intensive farming of the Iraqw and that of the Engaruka farmers as well as the present Sonjo (Sutton 1986: 48). In opposition to that view, Nurse and Rottland could identify no demonstrable Cushitic traces in Sonjo irrigation vocabulary. They also observed that the Sonjo vocabulary for irrigation features and practices remain semantically transparent, suggesting that the irrigation is 'likely to be of relatively recent origin' and set up *in situ* (Nurse and Rottland 1993). Therefore, in spite of various attempts to identify specific ethnic and linguistic connections, hard evidence is elusive. Communities practising intensive farming can be found within all the main language groups of eastern Africa (see Chapter 2).

THE SIEGE HYPOTHESIS

As a corollary to the common assumption that shifting cultivation was the norm in Africa, it has become common to explain its opposite, labour-intensive farming, as the result of situations of unusual stress, or 'siege', when people were forced to settle marginal land owing to political unrest. This sort of simplistic historical explanation runs through much of the writing on terracing in Africa. In West Africa, for example, Dogon is cited as a 'classic case of a harsh environment where people sought refuge from warring tribes' (Kassogué et al. 1996). Likewise, the stone walls at Maku in Nigeria are said to have come about when the transatlantic slave trade intensified, so that people fortified their villages and had to 'find ways to subsist on the rocky soil' (Igbokwe 1996: 223).

In the case of the terracing of Nyanga in Zimbabwe, Summers, despite the lack of dated evidence, vividly ascribed its origin to a situation

> when migrants or local people were driven to take refuge, with what little they could salvage, in the inhospitable mountains of the Inyangani range … Miserable and poverty-stricken in the extreme, the refugees eked out an existence on the cold and windy uplands for a few generations until the peril was overpast. Then once more they were able to creep down the sides of the mountains towards the lowlands, where conditions were more pleasant. (Summers 1958: 313)

It is notable that Soper, following his recent investigations in Nyanga, takes a different line. While political factors may, in his view, have played a role in the early part of the settlement, they 'do not appear to have been particularly pressing for the eighteenth century' (Soper 2000: 231). Moreover, and in contrast to Summers, he sees the fertility of the dolerite slopes as an opportunity for exploitation. Soper's conclusions point exactly to one of the shortcomings of the siege hypothesis: while a siege situation may well be one factor among others stimulating the settlement of these areas, it certainly cannot explain the persistence of the farming.

This idea of siege as a major cause for agricultural intensification was introduced by Gourou to the geographical literature in – apparently – later editions of his *Tropical World* (*Les pays tropicaux*).[13] Gourou was one of the geographers who saw cultural and historical, as opposed to environmental, factors as decisive in determining settlement patterns in tropical Africa. (See, for example, the references to Gourou's studies in Nigeria, Congo and Ruanda-Urundi in Trewartha and Zelinsky 1954: 151–2.)

In Gourou's later writings the siege hypothesis figured strongly. He cited (1991: 50) the Iraqw experience as a particularly clear example of intensive agricultural techniques developed by a people under siege. This siege thesis and its role in Iraqw historiography are scrutinized by Börjeson in Chapter 4. In the case of Marakwet, Östberg (Chapter 2) shows that the present insecurity and warfare in the Kerio valley have not stimulated intensification, but have rather led to the abandonment of some of the irrigated fields and hampered investments in new irrigation furrows.

It is of course consistent with the arguments opposing environmental determinism to consider, as Gourou does, settlement patterns and changes not as a direct effect of the environment. Indeed, numerous cases can be cited in the African literature of settlement moving either uphill or downhill in response to the local political situation rather than to population pressure or environmental changes. However, Gourou's emphasis on the siege hypothesis becomes almost as deterministic as the environmental determinism he opposes. The argument here is not that political factors never affected the development of intensive farming. The crucial problem with Gourou's use of the siege argument, rather, is the tendency to see intensification as something that comes about only under stress, to the exclusion of other factors.

An approach contrasting with Gourou's is that adopted by Kjekshus, who, in the words of Koponen (1988a: 241), 'summarily' dismisses the 'siege thesis'. Implicit in Kjekshus' argument (1996: 27ff.) is the assumption that the islands of locally developed, intensive agriculture are the remains of a formerly much more widespread agricultural regime. In his view the islands are pockets which have survived the perturbations that occurred in the wake of colonialism. In this interpretation colonialism was the main factor behind a disintensification of African agriculture. While there is little doubt that the advent of colonialism did indeed involve the disruption of local farming societies in a number of areas, it would be far too simple to advance this process as the main force behind a general and supposed disintensification of farming in eastern Africa. As Koponen has proposed, against Kjekshus, the migrations triggered by long-distance trade may even have indirectly induced the intensification of farming in some areas of Tanzania (Koponen 1988a: 240ff.).

THE SOCIAL CONTEXT OF LABOUR MOBILIZATION

From the above it should be clear that any explanation of intensive grain-based agriculture requires a broader understanding of the economic, social and historical contexts. Approaches towards such an analysis have been outlined by Hakansson (1989) who compares a subsistence-based model of agricultural development with one which, by contrast, focuses on pressure on production arising from needs for surplus. The demand for such surplus production might come from a chiefly elite. Commercial development, requiring increased production for a market, perhaps promoted by the elite in a situation of political stability, has also been considered as a stimulus for agricultural intensification (Hakansson 1989: 16).

The dynamic and uneven processes involved in the rise and fall of hierarchical political structures could indeed account for the emergence and decline of intensive agriculture, and also for its 'unpredictable' locational pattern. Furthermore, Hakansson argued, tribute labour controlled by chiefs and kings may be seen as one of the ways of mobilizing the labour needed for the large investments in land connected with this agriculture.

Two of the areas studied in this volume can, however, be used as arguments against hierarchical organization as a prerequisite for intensification (Widgren 2000: 262ff.). In the cases of Marakwet and Iraqw, labour, land rights and water rights are incorporated in a clan- and lineage-based social organization rather than controlled by a chiefly authority. In Marakwet, the mobilization of labour and the surveillance of the irrigation system rely on the decentralized power of the elders and are effected by common negotiations (Chapter 2). This also holds for the Iraqw (Börjeson 1999 and Chapter 4 in this volume). The empirical material thus shows that the mobilization of labour required for intensive agriculture does not necessarily require systems of tribute or other forms of social division. Labour can alternatively be organized through a system of age-sets or by mutual arrangements (such as work parties) within a more egalitarian, social structure.

The view that intensive agriculture may be associated with egalitarian social organization, rather than hierarchical control, is furthermore in agreement with Shipton's findings. In two papers he investigated the relation between, on the one hand, farming intensity and population density and, on the other, social structures in eastern Africa (Shipton 1984a, 1984b). Intensive farming in eastern Africa, in his analysis, is usually associated not with the centralized control of land, but rather with lineage- and clan-based land rights. In the field patterns this is reflected by a division of land into strips, 'an orderly pattern for lineage control of land' (Shipton 1984a: 618), whereas a more patchy system of fields is usually associated with chiefly control over land and less intensive farming. The Marakwet case clearly fits the strip pattern. Land there is held by clans and lineages (except in areas which have recently undergone government land registration). Furrows are owned by clans, and the genealogies, as perceived and acknowledged, provide a flexible system for dividing land currently under irrigation into strips. These property rights provide the stability needed for investments in land and at the same time the flexibility to cater for fluctuations in precipitation, as well as social and political changes. The crucial uniting point here, which Hakansson (1989) has clearly outlined, is the need for the social mobilization of labour.

The case of Konso (see Chapter 3) shows, however, that the dichotomy between a chiefly controlled tribute-labour system and clan-based acephalous forms of labour mobilization is less clearcut than either Hakansson (1989) or myself (Widgren 2000) have till now appreciated. The *poqallas* in Konso are hereditary male leaders, clan heads and also priests. But as Watson shows, they also act as big landowners and, in return for lending land and giving blessings and food, they receive labour.

Mobilization of labour is indeed an integral aspect of intensive farming in all the areas studied in this volume. Terraces, furrows and other structures require investments and repairs. Furthermore, with an increased number of crops per year, preparing the land, sowing, weeding and harvesting become potential bottlenecks. Tribute labour is one means of mobilizing the work force essential for this type of agriculture. But, as emerges from Watson's argument, different labour processes lend themselves to different kinds of social organization and are also situated in different social and cultural contexts. Our case studies show that traditional systems of labour exchange and/or work based on age-sets can provide such an input of work. Large systems of irrigation and field terracing do not necessarily indicate a hierarchical chiefdom structure. There is no simple relationship between the degree of agricultural intensity and the power structures involved. The crucial point is that all these communities have ways – but different ways – of mobilizing a concentrated labour input.

MARKETS VERSUS INSTITUTIONAL EXCHANGE

The stimulus of market opportunities, another argument brought forward by Hakansson (1989), has furthered agricultural intensification and associated investments in landesque capital in eastern Africa. One notable example in the nineteenth century was the emergence of irrigation farming by Lake Baringo (Anderson 1988, 1989) and its subsequent decay. More recently the well-known case of Machakos (Tiffen et al. 1994) must be seen in the context of its proximity to the market in Nairobi.

But this pattern does not apply in the areas described in this volume. Neither commercial market-driven networks nor caravan and urban demands (see Hakansson 1989: 18) explain their development. Indeed, several of these areas with intensive farming are nowadays relatively poor and located far from markets. In the case of the Iraqw in Tanzania one finds an actual inverse relationship between labour intensity and proximity to markets. In fact, the closer one is to markets in this area, and the more involved in market production the agriculture, the less labour-intensive it tends to be. Yet the labour-intensive core area excels in soil and water conservation but has poor roads and produces only a small share of cash crops (Loiske 1995, and Chapters 4 and 5 in this volume). Similarly, in Marakwet, the irrigated agriculture persists in spite of bad roads and poor market conditions; only a small amount of the produce is marketed (Watson et al. 1998, Chapter 2 in this volume).

In the Iraqw case, however, there is, as Loiske shows, a considerable amount of non-commercial exchange of agricultural products between the core and peripheral areas. Similarly Hakansson has recently illustrated the role of institutional exchange in precolonial production systems of eastern Africa, and especially the relation between such exchange and the intensification of agriculture (Hakansson 1994, 1995, 1998). The islands, as identified in this volume, are not – and have probably never been – closed entities relying solely on subsistence production. They rely on production and resource utilization from a range of economic and ecological zones. Exchange of products between different zones thus seems to be an important precondition for the existence of intensive agriculture. In the case of Iraqw'ar Da/aw, in the core area of the Iraqw, such exchanges take place within kin and the same ethnic group. In other cases, of which Marakwet is an example, exchange between cultivators and pastoralists of different ethnic backgrounds is an important incentive for specialization. This was formerly the case in Baringo (Anderson 1988, 1989) and may apply to a degree among the Sonjo, who are surrounded by Maasai (Adams et al. 1994).

Islands or icebergs?

The metaphor of an island to describe areas of intensive agriculture is adopted for the title of this book. It was first advanced by Maggs in a paper drawing comparisons at a broad level between Marakwet, a thriving agricultural community in Kenya (see Chapter 2), Ukara, an island in the literal sense with intensive agriculture in Lake Victoria, and Marateng, the remnants of a former agricultural community in South Africa (see above and Maggs 1995). At a first glance the examples treated here stand out as islands in contrast to surrounding dry lowland areas. Travelling to some of them may be compared with arriving at an island after some time at sea. The garden-like highlands of the Iraqw in Tanzania or the irrigated areas of the Marakwet in the dry Kerio Valley in Kenya do stand out from the adjacent dry savanna landscape. The merit of such a metaphor is that it encourages counter-argument. The papers presented in this book suggest that other metaphors might be equally appropriate. These islands are not in fact isolated, they are highly connected with their surrounding communities, not necessarily by market exchange, but through informal or institutionalized exchange, either within the same ethnic group or between ethnic groups and between different production systems. There are below the surface strong networks with the surrounding agro-pastoralist and pastoralist groups. What we see when we focus on the landscape and on the landesque capital might more appropriately be compared with the tips of large icebergs each with a vast invisible social landscape supporting it. This iceberg metaphor ought, moreover, to help us visualize the abandoned islands of Engaruka, Nyanga and Marateng in a new perspective. Whatever the reasons for their abandonment, it is more likely that their interaction with surrounding agricultural and pastoral communities will explain their dynamics rather than their assumed isolation.

Notes

1 This paper is developed from that presented at the World Archaeological Congress in January 1999 and at the conference on 'African Environments, past and present' (Widgren 2000). The research was financed by the Swedish International Development Authority (SIDA) and the Swedish Council for the Planning and Coordination of Research (FRN).

2 Intensive agriculture is in this context understood as 'farming systems based on short fallow rotations or permanent cultivation in which farmers employ labor-intensive techniques such as irrigation, manuring, and composting to maintain soil fertility' (Conelly and Chaiken 2000).

3 See the overviews by Ludwig 1968, Sutton 1984, Adams and Anderson 1988 and the special volume of *Azania* (24, 1989).

4 'Ich habe mich in Deutsch-Ostafrika monatelang in Gegenden aufgehalten, in denen ich während der ganzen Trockenzeit mein tägliches Bad in herrlich fließenden Wasser nehmen, meinen Durst unbesorgt, mit ungekochten Wasser stillen konnte, aber in der Nähe befanden sich nur ein paar ärmliche Hütten! Und auf Tagereisen war es nicht anders. Wasser war da, aber nur sehr spärliche Bevölkerung. Und wieder in anderen Gegenden, wo das Wasser tropfenweise in tiefen Felsspalten gesammelt, mühsam aus der Tiefe herausgeholt und dann noch auf steilem Bergpfad weit getragen werden musste, dort fand sich Dorf an Dorf und eine Bevölkerung mit Hochkultur.' (Nowack 1942: 368; English translation, MW)

5 Such a misconception is starkly illustrated by the following excerpts from a review of soil fertility research in a journal dedicated to understanding the dynamics of agro-ecosystems: 'Before colonization, and prior to the beginning of the 20th century, there was little agriculture in East and Southern Africa' (Nandwa and Bekunda 1998: 5); and later in the same paper, 'The introduction of continuous cropping partly coincided with the colonization of East and Central Africa' (p. 6).

6 Scoones's recent studies (1997a, 1997b) of the history of soil fertility management in Zimbabwe are notable exceptions.

7 Further examples of different trajectories are represented by intensification in Machakos (Tiffen et al. 1994), and abandonment of intensive techniques on Rusinga island (Conelly 1994).

8 The degree of agro-diversity in these farming systems is still insufficiently explored (see however, Conelly and Chaiken 2000).

9 For references to some of the early written documentation on Taita, Marakwet, Ukaguru and Kilimanjaro see Adams et al. 1994: 19; on Pokot, Hecklau 1970: 475; on Rusinga and Mfangano, Conelly 1994: 145 and on northern Kakamega in Kenya, Conelly and Chaiken 2000.

10 A similar approach to dating based on individual furrow genealogies and, moreover, on a relative sequence of superposition has been outlined by Soper for Marakwet (1983: 78).

11 See however Conelly and Chaiken (2000) for recent dietary surveys carried out in Kakamega district in Kenya, which questions too optimistic assumptions on the correlation between intensive agriculture and food security.

12 Ehret in his *Civilizations of Africa* (2002) deals with the same problem in a more generalized way. He puts a strong emphasis on the role of Cushitic-speakers in this development and also assumes very early dates of irrigation and terracing in north-eastern (second millennium BC, pp. 128–34) and eastern Africa ('well before 1500' AD).

13 Koponen (1988a: 241) refers to an edition of Gourou from 1966, which I have not been able to read, but in an early edition (Gourou 1948) there is no clear reference to this.

Two

The Expansion of Marakwet Hill-Furrow Irrigation in the Kerio Valley of Kenya[1]

WILHELM ÖSTBERG

It is a tricky operation to take water out of the rivers that hurl themselves down the very steep escarpment which drops from the Cherangany Hills into the Kerio Valley, more than a 1,000m below. Water is led through headworks, dams and sluices into kilometre upon kilometre of furrows, carefully levelled and embanked along the escarpment face and utilizing, where necessary, aqueducts made of hollow tree-trunks and shelves supported by wooden scaffolding along almost vertical cliffs, eventually to reach the valley and to irrigate the fields. The canals are constructed and bounded with rocks, boulders, logs and brushwood, reinforced by soil and grass. Nowadays, plastic pipes and concrete are important improvements or supplements in difficult places, but it is still local knowledge and local materials that keep the water running. About 40 channels are in use and they irrigate at least 4,000ha. (Mapping and descriptions are contained in Soper 1983 and in Watson et al. 1998.) Seeing that in Kenya as a whole the National Irrigation Board manages seven modern irrigation schemes covering in all little more than double that area, it is clear enough that the Marakwet people of the Kerio Valley have developed a significant resource over the years (Watson et al. 1998: 71ff.).

The Kerio Valley is a branch of the East African Rift Valley. A steep escarpment to the west rises about 1,500m from the valley floor. Irrigated agriculture has its main concentration in the Marakwet part of the valley, but is also found in Keyo areas to the south and around Lomut in West Pokot to the north. The Marakwet people inhabit the west of the Kerio River. Most of them live in the highland part of their area. (See Figure 2.1.)

The actual date when farmers here took the step from scratching a simple ditch in loose soil to embarking on the construction of the spectacular channels

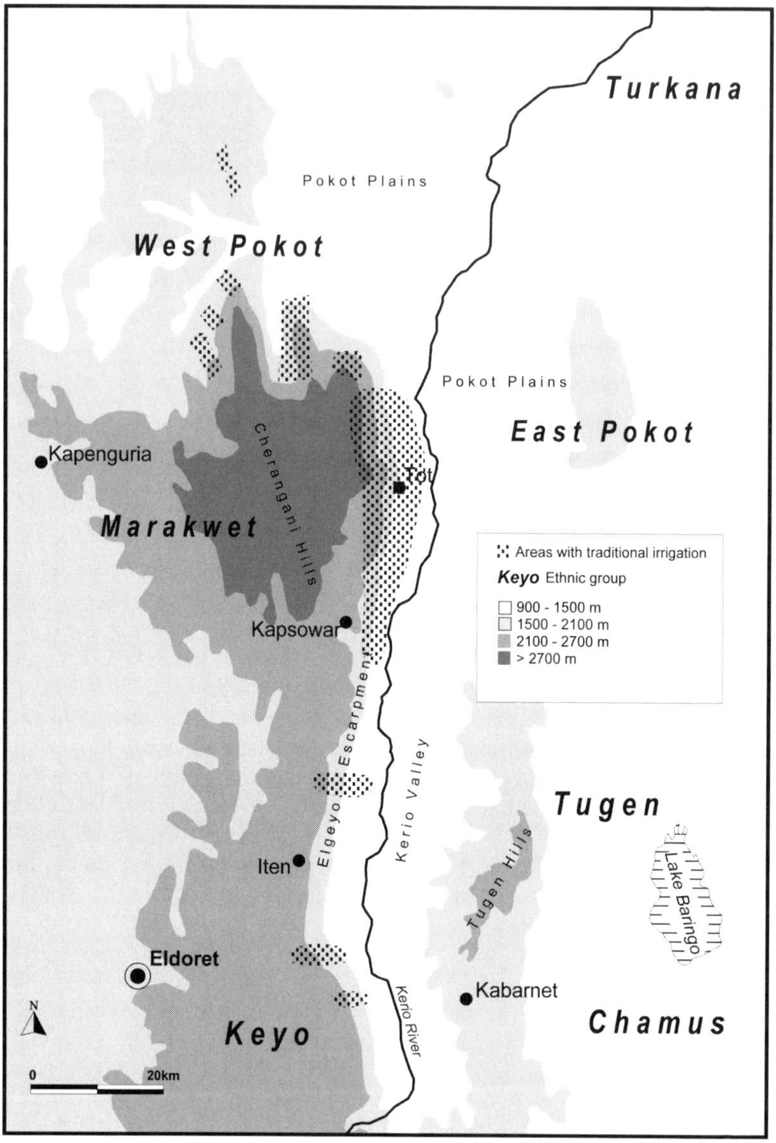

Figure 2.1 *The Kerio Valley (Map: K. Svae)*

on the escarpment can probably never be determined. But that should not prevent us from attempting to build a model of why and how it all happened. However, the expansion of hill-furrow irrigation in Marakwet is not exclusively a historical topic. New furrows are still being constructed, as we shall see, and others have been added within living memory.

Considering the engineering feats required to bring the water down the steep escarpment, this may be seen as a remarkable and indeed spectacular achievement. But, from another angle, it is something commonplace, a means of producing food. And this is how the Marakwet talk about their furrows. They have no words that can be directly translated as 'irrigation management'. They say merely *keir ber*, 'to water the land', and *kekwat ber*, meaning 'to level land' so that the water can reach all of it. This terminology illustrates how Marakwet irrigation is not the preserve of a class of specialists or rulers but constitutes a system with which most families are conversant and a resource on which they depend.

When the explorer Joseph Thomson, leading an expedition of the Royal Geographical Society through Maasailand in 1883–4, reached the Elgeyo Escarpment, he observed water channels, many of them 'conveyed with surprising judgement along the most unexpected places'. He noted, moreover, that irrigation here was conducted on a larger scale than in the Taita Hills in eastern Kenya. He was also to quick to learn that it was fully operative, for the water was promptly cut off from the camp site when his expedition was slow to pay a suitable fee for setting itself down in the area. 'We humbly paid up; and, immediately, as if a modern rod of Moses had struck the rock, the water began to flow' (Thomson 1887: 310).

From Thomson's account, it is clear that Marakwet irrigation furrows were in use well over 100 years ago. Similarly, in West Pokot, in the country immediately north of Marakwet, irrigation was observed in the same period by some of the first European travellers (Austin 1903, von Höhnel 1892). A considerably longer history is suggested by the technical similarities, at a broad level, between the furrows along this escarpment and the extinct irrigation system at Engaruka in northern Tanzania, abandoned some 300 years ago (see Chapter 6). Sutton comments that, if these two systems were related to each other, one has the attractive perspective of 'imagining an early extension along the Rift Valley of grain-cultivators with a locally evolved irrigation technology, well before the seventeenth and eighteenth centuries' (1990: 41, 1993). Ehret (2002: 394) insists, without going into details, that Marakwet irrigation is more than 500 years old.

Drawing on ethnographical and historical data and landscape studies, I shall attempt a scenario of how hill-furrow irrigation evolved in Marakwet. An understanding of its current operation is essential in this exercise, with the obvious caveat that one cannot simply infer the agricultural practices and social institutions of earlier periods from those at present or recently obtaining.[2] Moreover, little precise information is available on the Marakwet before the colonial period (H. Moore 1986: 14–15, Kipkorir 1973). Of necessity, the exercise must be conjectural in part and may lapse occasionally into an indeterminate, ethnographic present.

Approaches to agricultural intensification

Knowledge of local, hill-furrow irrigation in East Africa has been considerably improved in recent decades (for reviews see, for example, Adams 1989 and Adams et al. 1994). All the same, we still lack an empirically based understanding of how it evolved. With the exception of Sutton's work at Engaruka in Tanzanian Maasailand, no detailed, archaeological investigations have been undertaken at the various localities with irrigation, and no precise dating has been established for the furrows. We do not know what induced people to start constructing the furrows and when they did so. Taking a broader view of the issue in the context of the history of agricultural intensification in general, there are, however, a number of considerations which may inform our enquiry. This requires taking a detour through some varied lines of thinking, to identify those which promise to assist our quest and to eliminate those which seem invalid or inapplicable to our purpose. Following that, we may return to the Kerio Valley and delve into ethno-history.

1. *Evolution.* In an evolutionary perspective, it is conventionally imagined that societies proceed from hunting and gathering through pastoral and agro-pastoral stages to more intensive forms of cultivation. This line of reasoning is commonly embraced by agricultural specialists and development workers. In practice, however, one can cite numerous examples of societies, including instances in this very region, which have not followed this theoretical progression. By Lake Baringo in the Rift Valley, the Chamus during the nineteenth century combined foraging with farming irrigated by the Perkerra and Molo Rivers (Anderson 1988, 1989; Little 1985, 1992; Spencer 1998). At that time, one of the caravan routes from the coast passed through their area, so that the Chamus were able to sell their surplus produce in years with good rainfall and adequate irrigation water. However, the new income was largely invested in livestock, which over time led to a strain on the manpower needed to maintain the irrigated farming. Later, in the 1950s, some of the Chamus tried reverting to farming but with little success. Thus, this area, which in the nineteenth century had served as a granary for trading caravans from the coast, now struggles with widespread poverty and requires food relief from time to time.

The Chamus example, here condensed into a shorthand brief, demonstrates oscillation between different modes of subsistence in a way that defies neat assumptions about economic evolution. Indeed, such shifting balances, both within groups and between those with contrasting economic orientations, are frequent in East Africa and can involve change in or drifting of ethnic affiliation (Beidelman 1960, Muriuki 1974, Rigby 1985, Waller 1988, Spear 1993). Examples of the latter in the areas surrounding Marakwet are illustrated below.

2. *Population growth.* This is commonly evoked as the propelling force of agricultural intensification. While this school of thought has a long history, it received a new impetus from Boserup's classic study *The Conditions of Agricultural Growth* (1965). Her argument was that, as population increases and, as a consequence, the pressure on resources mounts, farmers gradually adopt more intensive forms of land use. But, since intensification is paid for by higher labour input, farmers will correspondingly lapse into more extensive modes of production as soon as the pressure on the resource base drops, notably when new land becomes available.

A whole literature has developed that discusses both the merits and the limitations of this framework. It has been suggested that conditions in the savanna areas of Africa do not fit the thesis very well. However, several recent studies in Africa inspired by Boserup's thesis have testified to its enduring importance, in particular by exploring the effects of high population densities on agricultural growth (Turner et al. 1993). Moreover, Netting in his seminal *Smallholders, Householders* (1993) acknowledged how his long-standing investigations of agricultural intensification in West Africa had been inspired throughout by Boserup's ideas. For Kenya similarly, the influential study of Machakos district, *More People, Less Erosion* (Tiffen et al. 1994), belongs in this category, although in this case the proximity to Nairobi makes developments as dependent on easy access to a huge market as on population growth.

The thesis that increased pressure on resources leads to intensification is, like any evolutionary theory, reasonable at the commonsense level but seems less incisive for explaining specific situations. And the process may vary from what Boserup's model might lead one to expect. For instance, in the Usambara and Pare Mountains of northern Tanzania, Hakansson identifies three separate instances in which Boserup's scheme does not fit the historical realities. Firstly, he demonstrates a case of intensive agriculture being abandoned, despite concomitant population growth (1989: 13ff.). Conversely, he points to a period of drastic depopulation in the Pare Mountains when irrigated farming was not abandoned in favour of more extensive modes of production (1995: 314). Thirdly, he found that high population densities in South Pare coexisted with large areas of unutilized land suitable for rain-fed agriculture (1995: 318). When we look at the history of the Kerio Valley, we shall see that irrigation gave rise to population concentrations rather than the other way round.

3. *Markets.* In recent studies, intensification is as often seen as resulting from market developments as much as from population growth, and the two explanations may be intertwined, as in the Machakos case just noted. Likewise, in the Baringo area in the nineteenth century, the expansion of irrigated agriculture was directly related to the caravan trade. Another example of how market demands facilitate intensification is the small town of Mto wa Mbu in northern Tanzania, where the tourist trade fuels the expansion of irrigated agriculture (Arens 1979, Braun 1990, Yanda and Mohamed 1990).

However, this explanation does not fit the situation in the Kerio Valley.

The irrigation furrows here predate markets of any size, and the Marakwet agricultural settlements were only very marginally on a caravan route. Even now, it remains difficult to transport agricultural produce efficiently from the valley to urban markets.

Nevertheless, if marketing in the modern sense does not explain the agricultural intensification in the Kerio Valley, exchange linkages between groups have indeed influenced the expansion of irrigated farming in Marakwet, as we shall see shortly. Another factor which encourages surplus production nowadays and induces people to keep flocks of small livestock is the demand created by ceremonies. Since many of these continue for days and involve many relatives, neighbours and visitors, considerable amounts of food and drink are required every year to keep the ritual life in good fettle. Thus, despite the absence of a commercial market stimulating agricultural intensification in the Kerio Valley, there would nevertheless have been substantial, local, economic exchanges and other social requirements generating a demand for surpluses.

4. *Centralization.* A proposition which was for a long time influential in world historical studies assumed that agricultural intensification was dependent on hierarchical social systems. Wittfogel's (1957) well-known hypothesis concerning oriental despotism, postulating a correlation between irrigation and despotic government, belongs in this category. Wittfogel noted (p. 166) the example of West Pokot irrigation, which constitutes in effect a northward extension of that of Marakwet.[3] Wittfogel's hypothesis, however, does not fit the realities of the Kerio Valley. Marakwet is an acephalously organized society, in which social differentiation is comparatively slight. Before the beginning of each cultivation season, people enter into prolonged local negotiations concerning the fields to be cultivated that year and how water is to be distributed. When these negotiations are concluded, both those with inherited primary rights to water and their less advantaged neighbours usually have access to land to cultivate and water to irrigate their crops. (See Ssennyonga 1983 and Adams et al. 1997 on 'ownership' of water and actual practice.) Since the irrigated area in Marakwet extends over at least 4,000ha, and the total length of the furrows about 250km, the management being decentralized and egalitarian, its challenge to centralization arguments is obvious enough. Indeed, in a recent attempt to resuscitate aspects of Wittfogel's thesis, Price (1994: 197) was forced to concede that a wide system administered in local units defies the model.

5. *Politics and the siege hypothesis.* Another view holds that political and military conditions necessitate intensified agriculture in particular areas. An example close at hand is the Mama Issara area of Mbulu District in Tanzania, where it has been assumed that Maasai raiding from the surrounding plains had forced people to develop intensive farming in the relatively secluded highlands. However, Börjeson and Loiske (Chapters 4 and 5) offer other more reasonable interpretations of how the intensive land use in Mama Issara evolved. Another

area of Tanzania, namely South Pare, has been similarly reinterpreted. There, it had been suggested that intensification resulted from confinement to small areas because of political unrest. However, Hakansson (1995: 312ff.) finds no historical evidence in Pare to support the siege hypothesis; rather, he argues the opposite, that of exchange relations propelling agricultural intensification.

While it could be argued that sudden political changes may have stimulated certain instances of intensification, this sort of explanation fails to account for how such production systems with their considerable demands for labour inputs have continued to thrive for generations. It clearly has no relevance to the Kerio Valley and the long history and development of Marakwet irrigation, as this chapter demonstrates.

6. *Ethnicity*. In some early published descriptions of irrigation in the Kerio Valley, as well as in accounts penned by colonial officers in the district records, one encounters the notion that a previous (and supposedly more intelligent) population had been responsible for the initial construction of the furrows. This reflects a deep-grained, historical outlook of that period, one which has been laid to rest by the scholarship of the postcolonial period. On taking a broader look at East Africa, one finds examples of intensive-farming systems which have arisen among diverse peoples and language groups. These include Kalenjin-speakers of the Nilotic family, in the case of the irrigation of Marakwet and the neighbouring Pokot and Keyo, Maasai-speaking Chamus at Lake Baringo, various Bantu-speakers, including both Sonjo and Chagga, and the Southern Cushitic-speaking Iraqw of Mama Issara, the last irrigating crops on valley bottoms but lacking hill-furrow irrigation. (For references see Figure 1.1.)

7. *Connectedness.* The foregoing survey has briefly examined six different approaches, which have from time to time been offered as explanations of agricultural intensification, none of which, however, appears particularly helpful for understanding the development of Marakwet irrigation. By contrast, I want to emphasize flexible adaptations, connecting mechanisms and comparative advantages. Bebbington (1997) has demonstrated how organizations and networks can make new resources and opportunities available and generate intensive modes of subsistence. The specific institutions and processes that he studied in six different localities in contemporary Ecuador and Bolivia have, of course, nothing in common with the situation in the Kerio Valley several hundred years ago. But the way in which he brings together a number of different dimensions of intensification (ecological niches, new products, technology, networks, brokers, etc) is akin to the discussion below of how irrigation evolved in the Kerio Valley. More generally, inspiration has also come from a number of recent studies that highlight the flexibility and experimental character of small-scale, African agriculture,[4] the adaptability of land management in Africa[5] and the resilience and dynamics of range-lands.[6] Instead of an evolutionary framework, one finds non-equilibrium systems in which resource use is a flux in time and

geography.[7] Moreover, the debate on the 'invention of tradition' school has been informative.[8]

To summarize the argument thus far, the expansion of irrigated farming in the Kerio Valley cannot be interpreted by single-cause explanations, be these population increase, market involvement, centralized leadership, specific historical events or ethnicity. Instead, it is proposed that the comparative advantages of different environments in and around the valley, in combination with the institutional arrangements of these furrow-making societies, facilitated the irrigation achievements. I shall try below to show that it is the dynamics of the local situation, and not particular events, that encouraged some people, on some occasions, to intensify their modes of subsistence. People found different ways of sustaining themselves. Over time, this resulted in a geographical division of labour, coupled with interdependence between groups. The overall picture is one of increasing variation rather than unidirectional development.

The Marakwet and their country

Nowadays, most Marakwet live in the highland part of Marakwet District, above the steep escarpment. Here, settlement reaches altitudes of 2,500 m. The area is cool and enjoys a high rainfall, in striking contrast to lower or eastern Marakwet, below the escarpment and bordering the Kerio River, which is the main concern of this chapter and where the average annual rainfall is in most areas well below 1000 mm (Figure 2.1) In view of the annual variability and the high evapo-transpiration in the hot valley, farming would be a high-risk undertaking if it were to rely on the rain alone. However, the several rivers, both permanent and seasonal, which descend the escarpment, facilitate irrigation at the base, thus rendering the western side of this stretch of the Kerio Valley favourable for habitation. The valley floor contains deep and reasonably fertile soils (Cappon 1985, Critchley 1979). Some cattle and small stock can be grazed. The area is rich in wild vegetables and fruits, and is also suitable for beehives. Formerly, it constituted good hunting ground. Recent armed clashes have, however, left large parts of the valley unutilized (see note 11).

ORIGINS

The Marakwet in the Kerio Valley have evolved from a number of groups who immigrated from various directions. Several of them mention Mount Elgon as a notable landmark. Some say that they moved through the lowlands, others through the highlands. Some claim that they came from the east and have memories of passing Mount Kenya even. These small groups were flexible and situation-specific, frequently encountering each other and separating, in no way resembling fixed ethnicities occupying set territories.[9] It would not have been possible to draw a tribal map of the Kerio Valley in the way that the colonial authorities did much later. This theme of fluidity recurs in the traditions of the other peoples now inhabiting and bordering the Kerio Valley, that is, the Keyo, the Pokot and the Tugen, situated to the south, north and east respectively of Marakwet. It is furthermore borne out by the close, linguistic relationship between these northeasterly, Kalenjin sections, with a dialect continuum from Marakwet in all these directions (Rottland 1983). Given today's distinct ethnic boundaries and that ethnic clashes are such a pervasive feature of life in the Kerio Valley at present, this atomistic background and the stories of fusion and fission are remarkable. However, it is this particular point, on which the current inhabitants of the valley insist, which forms the cornerstone of this interpretation of the expansion of irrigation. The upshot of this is that one cannot usefully search for a particular people who may have brought the irrigation techniques to the valley.

As the Marakwet see their history, in particular their early experiences in the Kerio Valley, their forefathers had livestock, settling temporarily to cultivate, joining with other small groups and splitting by turn, as particular families decided to move to new places. Disease or failure of crops could provoke movement, as could also attacks by cattle-raiders. In these stories of migrations and encounters, cooperation and competition, attacks and withdrawals, nothing particularly significant distinguishes one small group from another. It was a shifting way of living, with subsistence derived from livestock and cultivation, hunting and gathering. It was among those loosely structured, small groups that the irrigation works commenced.

EVOLVING ETHNICITIES

The Marakwet say that their forefathers met the now extinct Sirikwa people on the Elgeyo escarpment. The latter were primarily livestock-keepers who are known from the archaeological traces that they left across the Western Highlands of Kenya, a region which they dominated about 500 years ago (Sutton 1973, 1987, 1990, 1994). By around 1800, following the expansion of the Maasai, the Sirikwa were largely absorbed, partly into the Maasai, but more into the Kalenjin groups in the Western Highlands. Regarding the Kerio Valley, which appears not to have been exploited by the Sirikwa and their cattle, Sutton (1993: 57) suggests that at that period 'conjecturally … ex-Sirikwa elements moved down the escarpment and merged into the irrigation cultivators'.

The Marakwet remember clashes with the Samburu in the Kerio Valley, but they also tell of peaceful coexistence. By the middle of the nineteenth century the East Pokot had taken over as the major pastoral group in the Kerio Valley (Bollig 1990).

In the south, the Marakwet border on the closely related Keyo. An account published in 1927 emphasizes their pastoral lifestyle, and it is also noted in passing that irrigated agriculture is carried on (Massam 1927: 4, 111, 126ff., 134). This account is typical of how writers of the early twentieth century were biased towards emphasizing that the peoples of the Rift Valley were pastoralists at heart, although some of them did in fact practise irrigated agriculture. This is true also of the West Pokot, bordering on the Marakwet to the north and sharing many cultural, economic and social features with them and also making use of streams from the highlands to irrigate crops of millet and sorghum. In 1911, it was reported that in some years they were 'so poor and improvident that in times of drought they are reduced to feeding on berries, vermin and elephants' (Beech 1911: 11). But when a surplus was produced, it was sold to neighbouring pastoral peoples (ibid.: 17). The Pokot were assumed to have been originally a 'purely agricultural people' (p. 4) until some of them became pastoralists (pp. 4, 14), and in the 1930s it was assumed that they were increasingly moving into livestock-keeping (Maher 1935: 3). These brief excerpts from the early written accounts of the area, sweeping and imprecise as they are, nevertheless point to the fluid and changing character of production in the Kerio Valley and the adjoining highlands, as well as to trading and other links between the people living there. They bring out the role of livestock in the relations both within different groups and between them.

Further north are the pastoral Turkana. The contacts with them may largely have been a nineteenth-century phenomenon, when they expanded southwards (Lamphear 1988). From their area came goats, salt for the livestock and soda for preparing the much-appreciated snuff. The Marakwet offered tobacco in return. The goods thus obtained were not used in Marakwet exclusively but were traded further south (Massam 1927: 122). In the mid-1970s, I interviewed men in their seventies in the highland parts of Marakwet who, as young men, had embarked on trading expeditions to Turkana areas.

The East Pokot, across the Kerio River from the Marakwet, occupy drier areas than the Marakwet. They expanded rapidly during the first half of the nineteenth century, by integrating immigrants from several areas, including the Cherangany and Sekerr ranges and the Kalenjin groups living around Mount Elgon, as well as by successfully raiding their neighbours for livestock (Bollig 1990: 77). Before 1800 the Pokot are thought to have had a mixed economy, cultivating millet and sorghum and raising small livestock (Beech 1911: 4, Bollig 1990: 75, Sutton 1993: 57). Their pastoral specialization developed in the Kerio Valley. The East Pokot and the Marakwet have, throughout, been linked to each other by marriage, trade, celebration, warfare, child exchange and barter, but they have also been separated from each other by raiding and warfare.

Westward, looking up the escarpment, the Cherangany Hills were covered by deep forests, that is before the agricultural settlement and clearance of the second half of the twentieth century. Cultivated fields could be found along the fringes of the forest, which was also used for grazing animals, for hunting and for placing beehives but not for permanent settlement. Scattered in the hills were small communities of Kalenjin-speaking Okiek, who subsisted by hunting and gathering and who were also craftsmen. The immense forests, as well as the risk of raids by the Uasin Gishu Maasai and also by the Nandi who in the mid-nineteenth century were expanding northward on the Uasin

Gishu plateau, deterred the Marakwet from establishing themselves permanently in the highlands.

COOPERATION AND HOSTILITIES

Nowadays the economic differentiation between the peoples inhabiting the Kerio Valley may appear obvious enough. Marakwet irrigated farming constitutes one extreme and East Pokot pastoralism the other. However, their modes of subsistence are not, and were not, as clearcut as the terms 'pastoralists' and 'irrigation farmers' may suggest. The irrigation farmers have kept some livestock, and the pastoralists have cultivated grains.[10] There have also been intermediate categories of those who combined less sizeable herds with farming on a modest scale where irrigation water was not available, as in the Tugen Hills. The balance between the different methods of sustenance might change from year to year, depending on the variable rainfall and the incidence of pests and diseases, and nowadays as much on armed attacks, variable markets and unpredictable aid projects.

The various groups living in the Kerio Valley have exchanged grains, livestock and other goods with each other as far back as memory goes, and this has continued into modern times. In 1911, Beech noted that Pokot agriculturalists traded handicraft articles against sheep and goats with the pastoralists (p. 17). He also noted that grains were part of the daily diet of the pastoral East Pokot and that they could be obtained from the Marakwet. When Major Austin in 1897 obtained a guide in West Pokot to assist his advance towards Karamoja and Turkana, this person knew the languages of those areas and his 'intimate knowledge of the country traversed was extraordinary' (1973: 62). Traders from afar also touched on the area. In the Baringo area, coastal ivory-traders are reported to have appeared around 1840 (Spencer 1998: 148). Perhaps this was also the time when traders ventured into the Kerio Valley looking for livestock, hides and skins, and ivory. From the mid-nineteenth century, the caravans travelling the northern leg between the coast and Lake Victoria passed by Lake Baringo, over the Tugen Hills and on to the Uasin Gishu Plateau often through Mutei in Keyo (Sutton 1973: 85ff.). The explorer Joseph Thomson's account of how in 1883 he was charged for setting up his camp (see above) and punished when he was slow to pay up suggests that when the first Europeans reached the area, the villagers had clear ideas about how to deal with outsiders.

In the 1920s, the Keyo obtained pots from Marakwet areas (Massam 1927: 120). They were exchanged at markets, as well as obtained through interpersonal networks. In exchange for milk and meat, pastoralists got grains, spearheads, cowbells, iron ornaments, knives, pots, gourds, tobacco and other goods from the Marakwet. Honey was produced by all groups and exchanged according to demand and supply. In Kalenjin societies, honey beer is an important commodity. Large quantities were, and are, required for the major rituals. It also enters into marriage negotiations and thus was, and is, an important item of local exchange. During a study of the material culture of

Baringo District, bordering Marakwet to the east, Hodder found pots and spears from the Tot area of Marakwet in a number of homesteads north and east of Lake Baringo some 80 km away (1982: 39, 64). On juxtaposing scattered notes on trading relations, as I have now done, a pattern of widespread exchange among the peoples inhabiting the Kerio Valley and the surrounding hills emerges.

There is, furthermore, a long history of marriage relations between the peoples living in and around the valley. Hodder shows in instructive maps that most people in the border areas of the Pokot and Tugen have in-laws in other ethnic groups (1982: 23ff.). From the Chesegon area, Kurita reports that between one-fifth and one-quarter respectively of the households on both sides of the Marakwet-Pokot border had wives from the opposite group (1983: 66). It was also common to have 'cattle friends' across the ethnic boundaries. A livestock owner would leave some of his animals with partners in various parts of the area. The partner could use the milk and also keep a share of the progeny. The livestock owner would be safeguarded against localized disease and raiding and could in times of crisis count on support from his 'cattle friends'. Hodder found that it was common to have 'stock friends' across ethnic borders in the Lake Baringo area (1982: 23ff.) This was likewise a practice between the Marakwet and the Pokot, but, given the current hostilities in the Kerio Valley, the situation is different now.[11] Bollig writes that, among the East Pokot, a livestock owner maintained 'stock friends' in between six and ten different places but that it was rare to have two partners in one place (1998: 142).

Numerous families were thus linked in enduring and important relationships to people across what are now regarded as distinct borders. When droughts and other afflictions befell them, they sought refuge among each other. Oral histories around the Kerio Valley abound in accounts of how famine and raids forced different groups to move and settle with neighbouring peoples (Massam 1927: 13ff.). Among the Marakwet, different descent groups have their origins among the Keyo, Chamus, Luyia, Maasai, Okiek, Pokot, Tugen or Turkana, just as Marakwet also have their kinsmen scattered among other peoples.

The Marakwet grazed their livestock along the Kerio River, and they placed beehives on its eastern side in Pokot and Tugen territory. Occasionally, thefts of honey stirred up feelings, but participating in each other's celebrations strengthened more amicable relations. Concerning the Kiptani, a Keyo group living immediately south of the Marakwet, an early account described their country as consisting of 'nothing but rock. Not a blade of grass grows there except during the heavy rains. Fortunately, the Kiptani are, as a rule, very friendly with their neighbours to the north and with the people across the river … share grazing on the less arid country' (Massam 1927: 18). Further south, the Keyo grazed their goats together with the Tugen (Massam 1927: 43). At the northern extreme of Marakwet the Pokot at Chesegon have used water originating on the Marakwet side for a very long time, a practice which has, however, been increasingly challenged recently. When the West Pokot district commissioner in 1959 removed Marakwet from one part of the area, those affected by the move retaliated by closing the water furrows going towards Pokot.[12] Again, after independence, when the border at Chesegon was redrawn to the detriment of the Marakwet, they cut off the water once more. The serious clashes in the last few years have led to the same result.

Nowhere in the district is the ethnic boundary between the Marakwet and the Pokot nowadays as clearly shown in dress and in material culture as at the River Chesegon. At this created boundary, it has obviously become particularly important to indicate to which side one belongs.

Men who are now in their 70s and 80s relate with enthusiasm how they herded together with the Pokot along the Kerio River, how they feasted together and how some of them settled on the other side and became heads of new lineages. Some lived there for a couple of years and later returned. Many Marakwet livestock keepers underwent the *sapana* ceremony and adopted Pokot attire.[13] Until the severe hostilities of the last few years, a number of Pokot youth came for circumcision on the Marakwet side, one result of which was to draw some of their Marakwet age-mates towards the life of the plains. North of the Marakwet, it has been common among the East and West Pokot to oscillate between the different economies of the plains and the hills (Dietz 1987: 160ff., Edgerton 1971: 82), although, more recently, population increase and land adjudication have made it increasingly difficult to uphold such flexible adaptations to changing conditions (Dietz 1987, *passim*; Nyberg 1999: 65ff.).

It is not being implied that people are free to move between different places, economies and ethnic groups according to whatever whim takes them. Most people live their lives where they were born. Nevertheless, in the Kerio Valley, one frequently encounters individuals and families who have broken away from their established mode of life. The various groups inhabiting the valley have had knowledge of and a need for each other's skills and resources, and numerous relationships have been devised. In Marakwet agriculture intensified in an environment characterized by widespread and multipurpose relationships.

If interaction and cooperation have been common, so also have raids and confrontations. It is a recurrent theme in local accounts of Marakwet history that raiding and warring were much more common before the arrival of administration in the area in the 1910s. A cattle raid or a theft of honey from a beehive could easily be followed by counter-attacks, further raids and warring. In such situations, markets were closed, visits to relatives across the border became impossible and grazing agreements became void. Such short-term crises were common during the nineteenth century, and probably also in earlier times, according to what elders related in the mid-1970s. At that time, this situation still prevailed, although skirmishes never developed into warring. The few old rifles kept at the administrative police posts were sufficient at the time to guarantee peace in the valley. Twenty years later, the situation had changed completely. Throughout the 1990s, cattle thefts and attacks on villages occurred in the Kerio Valley. There are no longer any cultivated fields along the Kerio River, and those close to the escarpment in the west are cultivated under armed guard by young villagers carrying automatic weapons. The settlements in the valley are guarded day and night against attacks.

As farmers have been – and still are being – forced to flee their communities for shorter or longer periods, obviously the irrigation furrows cannot be properly maintained. Problems of soil erosion or declining soil fertility are trifles compared with the threat to irrigated farming in the Kerio Valley posed by armed conflict. Large numbers of farmers have now left the valley. Most schools have closed. The cultivated area has been diminishing year by year.

One further serious consequence of the large-scale attacks is that, as people are forced to move to the highlands, the forests of the Cherangany Hills, which currently constitute a protected water catchment area, will unavoidably be cleared for settlement, with disastrous consequences for the valley below. There is reason to fear that the permanent rivers that feed the water furrows will fail during the dry season.

Looking back at the history of the Kerio Valley, one finds both fluidity between the groups and the creation of specific identities. Those who wanted to monopolize particular environments needed the support of others' labour to exploit them. There have been periods of incorporation and periods of exclusion. Lately, the valley has experienced a phase of extreme polarization.

It had been possible to cross borders to use resources not essential to those inhabiting a particular area. But herds had to be kept away from growing crops, just as fields could not be allowed to encroach indiscriminately on pastures and watering-places. In such ways, the groups created space for each other's activities. Thereby the valley can be said to have grown bigger, as it was utilized for more specialized undertakings, widening the spheres of subsistence. The Kerio Valley was not an expanse of endless resources but neither was it a fixed larder. The expansion of irrigation is an obvious illustration of this condition.

The beginning of artificial irrigation

Having sketched the historical background, I now draw on local, social and cultural considerations to offer suggestions on how irrigation in Marakwet evolved and the system developed.

Constructing the hill furrows would from the start have been a form of specialization, but one very different from the yield maximization kind of farming commonly associated with modern irrigation (Adams 1989: 26). It was part of a subsistence system, and its principal value was that it made people somewhat less dependent on the irregular rains. Livestock, beehives and rain-fed farming continued to be important, as before.

Once a few people had made the attempt to assist their cultivation by leading water to the fields, they were naturally committed to that particular locality. Previously, in the Kerio Valley, the normal pattern would have been to adapt according to conditions and shift when necessary; now, with the beginnings of irrigation works, some people settled permanently, growing crops for subsistence as well as for barter. Larger families could be raised, crops could be stored more easily and crafts could be developed more effectively. Gradually the landscape changed. Along the permanent rivers which descend the escarpment, the Arror and the Embobut, more and more channels were constructed. Such investments had the effect of stabilizing the settlements and tying people to each other and to their chosen sites. Those who were to become the Marakwet began to distinguish themselves from other groups.

The channels were constructed by landholding descent groups. Since water distribution followed the same principles as land rights, labour was available when repairs to the works were needed. In order to get water to one's piece of land, one had to contribute to the regular upkeep of the channel.

No furrow bureaucracy was needed. All the interested members of a local community became involved, and the manpower required could be recruited. In principle, the rules are unambiguous: either one has or does not have a share in a certain furrow, and this depends on one's genealogical position. Access to the water is thus guaranteed to those who honour their obligation to participate in the necessary communal works to keep the furrows in good repair. Outsiders could be invited to join if they contributed their labour in constructing the canals. Those who did not join remained excluded from the water, a rule which obtains to this day, at least in principle: there are continuous negotiations over where to take the water in any given year, and also from day to day.

Talking to the furrow-makers, there is no doubt that they are proud of their achievements. They put in considerable amounts of work and they delight at seeing the job being done. Their effort is more than mere toil and drudgery. A working group in full swing is a festival of energy and rhythm (Richards 1986: 87, 1989). In Marakwet, as elsewhere, some people respect work and enjoy it, even when it comes to heavy farm labour.[14]

However, in much of the regional literature of the twentieth century, the local commitment to the irrigation system and the maintenance of the furrow works was barely appreciated. Among the West Pokot irrigation farmers, Beech noted that it was the 'aim and ambition of the hill Suk [that is the West Pokot] to amass sufficient livestock to enable them to descend into the plains and join the pastoral Suk' (1911: 4). A similar opinion was reiterated in 1979 by the anthropologist Schneider: 'In a land containing both good grazing and good agricultural land, they choose to grow crops extensively only if their cattle operations fall below what is economically supportable, at which point individuals move to the mountains and engage in agriculture until they can recoup their losses by selling grain to pastoralists' (1979: 85).

Massam (1927: 4) was of the opinion that the Keyo looked 'on work as an unmitigated evil … spend most of their time idly herding stock and making themselves ornamental'. Another district commissioner wrote in 1935 about people in two Marakwet locations which had irrigated agriculture that 'Endo and Mukorro are the most reluctant to go out to do

Figure 2.2 *Repairing a furrow in the highlands. Breakages may occur a long way from the irrigated fields in the valley, and demand that people promptly move up the hills for repairs. Before each cultivation season the furrows must be inspected and repaired.* (*Photograph: Mats Widgren, 1998*)

work'.[15] Fifteen years later, yet another colonial officer wrote that 'the Marakwet are not keen and efficient agriculturalists ... the hearts of almost all of them are not in the work of cultivation. Deep down they regard agriculture as an unfortunate necessity forced upon them against their inclinations by an unkindly God. Their better inclinations are all for the life of a stock-owner.' (Hennings 1951: 209)

I am not arguing against the opinions of these observers, and their number could easily be extended. These writers encountered the opinions and circumstances that they report. However, the opposite view could equally easily be recorded. This is, once more, the heart of the matter: people have made different choices, and economic and political forces have pushed people in different directions. In time, both furrow-makers and pastoralists emerged in the Kerio Valley. This specialization proved a favourable environment for expanding the furrows.

EXPANDING THE FURROWS

The first furrows would doubtless have been led off the lower reaches of the rivers and have been pioneered by small, local groups. But the construction of successive new furrows higher up the rivers would have been a very different

Figure 2.3 *Constructing a new furrow. The Talai clan of the Sambalat area in northern Marakwet is building a new furrow to take water 13km from the intake off the Embobut River to their fields on the valley floor. Samwel Kalarewei, chairman of the furrow committee, stands in the canal as it cuts its way through the escarpment.* (*Photograph: Wilhelm Östberg, 1998*)

undertaking. It would have required competent surveying, a large work force and a commitment to see the project through over many months, as well as resources to overcome the technical and social obstacles as they appeared along the route. The constructors had to foresee where water could pass on completion of months of communal work through uncleared mountain bush, with rocky screes, cliffs, gullies and gorges to be crossed. And they would have to persuade their neighbours to share the burden of the work, but equally the real risk of failure. Such grand schemes are still embarked upon. Soper (1983: 78) lists certain furrows constructed within living memory, and others have been added since his survey.

A yet more recent example deserves recording here. On 27 July 1996, a group of 22 people assembled for the first working day in the construction of a furrow from the Embobut River, intended to carry water 13km to the fields of the Talai clan section at Sambalat, just south of Chesegon. A previous attempt had been made in 1974 but had been abandoned after, as it is recorded, three landslides had occurred along the course being dug. The soil had been too loose to retain the water flow. This time, the Talai clan was prepared to try again. Until now, they had relied on seasonal streams, and they had also negotiated each year with other groups for additional water. But with a growing population, they felt the need for their own water. Apart

Figure 2.4 *A furrow moving gently along the contour through the forest high above the valley floor. A log of wood provides a simple foot-bridge so that people may conveniently pass without damaging the bank.* (*Photograph: Lowe Börjeson, 1998*)

from irrigating their land, the new channel would serve four schools, two churches, one dispensary and a trading centre, and would provide all the households along the route with domestic water. Livestock would benefit too.

Five elders had surveyed the route for the new furrow. When they embarked on the project, people said that they had gone for marriage negotiations. 'When water comes home, it is marriage. Whoever has a channel has a wife.' Owning a furrow is compared with having livestock: it attracts women. Young men in Sambalat joke over their prospects once the water has arrived: 'You will not need to talk long to a girl before she agrees. Women want a place with water, where mangoes grow.'

After two and a half years, this Talai group had taken the furrow through the mountain forest, had successfully crossed two gorges and had negotiated steep rock faces. The furrow passes through neighbouring villages and has reached their own location. It is still some way from the fields on the valley floor, but the conclusion of the project is clearly in sight. By November 1998 the size of the furrow work force had grown from the original 22 members to 98. Records are kept of attendance. Work proceeds on one or two days a week for long periods of the year. During the school holidays, when teenagers can join, some 50–60 people may turn up with *pangas* (machetes), hoes,

Figure 2.5 *Water reaching the fields on the valley floor. From this furrow water will be diverted into several rivulets and eventually the small rills watering the crops. (Photograph: Wilhelm Östberg, 1998)*

mattocks and spades to push the channel forward. But work has also had to be called off for months, following attacks by Pokot. The area has suffered losses of life, houses have been burnt and livestock driven away.

At an early stage of the project, the district water engineer paid a visit and endorsed the plan, confirming that the route surveyed was feasible and also calculating the costs of materials needed. Although the villagers have received occasional small contributions from outsiders to support the work, in all essential respects this new furrow is being built by local effort.

The people who in July 1996 embarked on the construction of a new furrow, more than a dozen kilometres away from their fields in the Kerio Valley, needed the water and were prepared to move it through gorges and past rocks, through other clans' territories and through deep forests. Not all of the villagers who might benefit were convinced at the beginning. But in time they have joined the project, as it has made progress.

COMMON DECISIONS

Irrigation in the Kerio Valley started piecemeal, as needs and opportunities occurred. But there have also been major projects, and these required a social

Figure 2.6 *A hollowed tree-trunk acting as an aqueduct in the Tot area, northern Marakwet, allowing the furrows of two different clans to cross.* (*Photograph: Wilhelm Östberg, 1974*)

organization able to undertake them. The Marakwet had this. As already mentioned, land was held collectively by kinship groups and allocated according to descent, in which all had their positions and stood to benefit. Another factor was the readiness to invite others to use land and water in exchange for their labour. Yet another crucial factor was the possibility of organizing labour via the age-grades.

The mode of organizing labour, arranging a marriage ceremony, clearing new land, resolving a conflict and putting up defences against raiders follows the same pattern. Whether enlisting the cooperation of only a handful of people, 50 or 200, the same procedure is followed. People meet and discuss and reason, without any chairman or judge being appointed. They listen to evidence, compare the case with previous cases, quote proverbs, recount particular events and reason again. This decision-making process is useful in solving small, everyday issues, such as which group of people should dig a specific stretch, or who takes the responsibility for crushing stones to build a reinforcing wall, or who repairs a broken aqueduct. It may also extend to decisions on constructing a new furrow. The extraordinary achievement of leading water along the steep escarpment face has not required a centralized political body or a dominant class or elaborate planning charts. It is this same pattern which allows irrigation to continue expanding today.

The Marakwet's immediate neighbours, the Keyo and the West Pokot, organize furrow maintenance and irrigation in ways comparable with those of various Marakwet sections. Other East African communities organize the

Figure 2.7 *Leading water off the main furrow to the fields of a group of farmers who are about to irrigate.* (*Photograph: Wilhelm Östberg, 1998*)

Figure 2.8 *Taking the water to the land.* (*Photograph: Wilhelm Östberg, 1998*)

Figure 2.9 *Water entering the field as a group of women plant finger-millet. The land has been cleared but the soil is not turned until the first weeding.* (*Photograph: Wilhelm Östberg, 1998*)

work in different ways. Across the Kerio River, among the Chamus, irrigation has been controlled by ritually strong councils linked to the age organization (Spencer, 1998). Among the Sonjo of northern Tanzania, water has been managed by a village council of ruling elders who reckon direct descent from those who were once ordained in these positions by Khambageu, a cultural hero of the Sonjo whose position equals that of God. 'These elders are believed to understand the irrigation system better than other people … the irrigation system would collapse in chaos if it were not strictly regulated' (Gray 1963: 143). This is a far cry from the negotiations taking place during the local water meetings in Marakwet, in which all members of a descent group participate.[16] Both the Chagga and Pare of northern Tanzania are people with chiefdoms, which produce lines of command and networks and alliances for managing land and water that are quite different from those of the Marakwet (Hakansson 1995, 1998; S.F. Moore 1986; Tagseth 2000).

MORE PEOPLE, MORE PARTNERS, MORE FURROWS

The more numerous the Marakwet became, the better use they could make of the land, the forests and the water and the less vulnerable they became to wildlife and to hostile raids. As in many other societies in Africa, the proverbs and sayings of the Marakwet emphasize that it is a good thing to multiply and to spread. In blessings, young people are encouraged to marry, to have children and to prosper. Polygyny ensures that as many women of fertile age as possible are married. Marriage should not only increase numbers but should also ensure that networks grow. Taboos against marriages between different categories of kin require that one finds spouses outside one's own immediate settlement, or indeed among neighbouring peoples. Interestingly enough, two seemingly contradictory principles are at work, because it is also considered advisable to marry into lineages where unions already established have proved to be stable and fertile. However, the contradiction may be more apparent than real. Together, the two principles both build and strengthen networks. Marriage is a means of easing access to water and land and to food in times of need.

The relatively secure production of food from irrigated fields attracted immigrants, as still happens. The insecure, natural and political conditions in and around the valley have always produced people needing relief. At times, people who had access to irrigated land could offer this. And extra hands made it easier to maintain and extend the furrows. Since there were clearcut rules about rights to water and land, and well-established routines for sharing water and negotiating temporary access to it and to land, outsiders could be invited without upsetting the normal course of events. As communities grew, this in turn led to a need to extend furrows and irrigated fields. The number of furrows expanded, as did the number of people. The process continually reinforced itself.

When the colonial administration was established in Marakwet in 1911, overt hostilities between the different groups in the valley were suppressed, which might have improved the chances of expanding the number of furrows further. It also became easier to penetrate the highlands. Iron implements assisted in the clearing of the big forest trees. The Maasai no longer controlled the Uasin Gishu Plateau, and in 1906 the Nandi had been restricted to their reserve further south. The Marakwet could graze their animals in the pastures above the escarpment in a much safer situation. Some took to rain-fed agriculture in the highlands. A new economy emerged above the escarpment, with different crops, a different agricultural calendar and access to new markets. The irrigation farmers acquired one more economy with which to cooperate. When crops failed in the lowlands, there were now new possibilities through markets, relatives and acquaintances in the highlands. This remains the case. Along the roads and paths connecting the highlands and lowlands in Marakwet, there is a steady traffic in maize and beans descending into the valley and in fruit and sweet potatoes being carried up.

As the number of irrigation channels increased, agreements had to be made between neighbours so that water could pass through areas belonging to others. Cooperation became necessary. Conditions were stabilized.[17] Moreover, people living in the drier parts of the valley benefitted indirectly from the increased reliability of grain production under irrigation.

Specialization in livestock-keeping could be extended since 'other' food was available when droughts and pests struck.

In recording the history of a number of the older furrows, stories were encountered of irregular inheritances and outright deceit, together with the expected accounts of descent groups having organized themselves to construct the furrows that they still use. Considering how vitally important rights to the water are, one would expect the owners of the water to parade the most respectable charters. While that is certainly the norm, stories of homicide, inheritance through daughters, fraud and magic are also narrated. It is as though their very impudence would provide them with stronger proof of an impregnable right to the water. The collective message of these stories of irregularities appears to mediate between, on the one hand, the clearcut rules defining who is a water-owner and, on the other, the de facto acquisition of rights to water by later arrivals in the irrigable areas. By accommodating accounts of water rights acquired by irregular or dubious means, even if with derision and head-shaking, it becomes possible in practice to uphold the moral and legal system regulating the control of land and water and at the same time to legitimize historical facts which seem to run contrary to the norms by which water is owned and utilized.

Attempts are now and then made to exclude some of the existing users from the water, causing them to complain. In two recent cases, the local court found that the complaining groups did not in fact have formal rights to the water. Yet it was ruled that all parties should continue sharing it 'as brothers'. Obviously, the consequences of depriving people of the water that they were using appeared to be too serious.

Cows, exchange and separation

Having outlined the conditions that promoted the expansion of Marakwet irrigation, so that eventually it reached its present extent and complexity, there remain two factors, already hinted at, which deserve more specific attention. These are the exchange of cattle and the trade in goods, both of which have been shown in other parts of East Africa to have influenced the evolution of intensive forms of agriculture.

THE ROLE OF LIVESTOCK

The irrigation farmers counted – and still count – on holding at least some grain in store in most years. But there are limits to what can be saved for poor years. Even under settled conditions, insects invade stores; therefore surplus grain is better used for acquiring animals to get milk and meat and as a precaution against drought; stock also serve in effect as capital and can be used to build up reciprocal relationships. In these ways they constitute an important source of security for individual households in semi-arid environments (see for instance, McCown et al. 1979: 299).

Livestock are moreover – as they have always been – a convenient form of investment. They are available locally, and extra animals can be added to the herd with only minimal extra inputs of labour. During the dry season, they can be fed on crop residues. Acquiring herds and building irrigation systems can thus proceed hand in hand as complementary pursuits. (For other East African examples, see Anderson 1988, 1989, Hakansson 1994, 1995, Waller 1985, Sutton 1990, 1993.) However, in Marakwet acquiring a large herd of cattle is not the all-important goal of every household that it is in several other East African societies. Here it is goats and sheep that are required for marriage negotiations, for exchanging for water rights, for ceremonies and the treatment of sickness, for paying fines and for slaughtering for visitors, as well, of course, for feeding one's family and for immediate cash needs. In the first place, it has been small stock that matter in Marakwet; the same applies in Keyo and West Pokot. Around Lake Baringo likewise it has been argued (Spencer 1998: 150ff.) that the irrigation farmers concentrated on only small stock until about 1865, when Samburu immigration to the area stimulated a growing interest in cattle among the Chamus. Previously, it seems, they eschewed the keeping of cattle, owing to the risk of being raided by pastoralists. This same point is made nowadays by Marakwet in the Kerio Valley, who aver that, if there were only goats and sheep in their area, they would present a less tempting target for armed assaults by the Pokot. Investing in livestock was, all the same, one logical form of using a surplus, and obviously some successful farmers would gradually move in the direction of a livestock-based economy. Some crossed the River Kerio and established themselves on the other side, becoming Pokot. In either case, they became trading partners of the irrigation farmers.

The pastoralists, for their part, were dependent on obtaining grain. On the basis of a comprehensive survey of production data from herds in Africa and the Middle East, and on the modelling of herd growth under several different assumptions, Dahl and Hjort (1976) show that a 'reference family' of two adults and four children would need a herd of 60 cattle to provide enough fresh milk for their needs and for the reproduction of the herd. During the dry season, when few animals lactate and the yields drop drastically, the required herd size grows tenfold (p. 158). Small wonder that their ethnographic survey confirms that 'pure' pastoralists include agricultural products in their diet.

In a similar vein, Schneider, who studied the livestock economy of the Pokot, has argued that 'any close examination of any of East Africa's pastoral people will show that they depend in varying degrees on consumption of grain and other foods' (1979: 64). Dietz (1987: 127ff.) provides data on conversion rates between livestock products and grains, demonstrating that, throughout the twentieth century Pokot pastoralists had favourable terms of trade with their cultivating neighbours. They had an incentive to trade, and the agriculturalists obtained an outlet for their surplus, which meant improved food security. As the trade went on, the furrows were expanded.

Although cattle ownership cannot be said to have been the decisive factor behind the intensification of agriculture in Marakwet, the proximity to cattle-based economies was important for its development.

THE IMPORTANCE OF EXCHANGE

In a set of articles, Hakansson has demonstrated how regional exchange systems characterized the precolonial economies of western Kenya and northeastern Tanzania (1994, 1995, 1998). Ambler has documented similar conditions in central Kenya (1988). Analysing the precolonial economic relations between the Luo and the Gusii in western Kenya, Hakansson argues that the demand for cattle as bridewealth payments sustained a regional exchange system, which in turn instigated agricultural intensification. The observation that the accumulation of cattle was linked to economic and social stratification is important in his analysis: 'The demand for cattle was practically unlimited and therefore geared Gusii households towards production for exchange' (1994: 271).

Similarly, he demonstrates how in the Pare chiefdoms in northern Tanzania, regional exchange helped to generate agricultural intensification (1995, 1998). The exchange of cattle and grains fuelled a quest for surpluses above household subsistence needs, which in turn promoted agricultural intensification. The caravan trade added further impetus, but irrigated agriculture had in fact long predated the caravans. In Hakansson's analyses, exchange relations and cloth, beads and other trade items underpinned agricultural intensification. In Pare, it is, he argues, 'difficult to reconcile subsistence needs as the basis for irrigation with the fact that substantial amounts of agricultural products were regularly exchanged for nonfood items' (1995: 317).

In the Kerio Valley, the situation was not quite the same. I have tried to show that irrigation evolved as an attempt to become less dependent on the unpredictable and insufficient rainfall. The expansion of hill-furrow irrigation was linked both to increased population density, facilitated by irrigation, and to the system of exchange that developed between different economies, as these emerged in a process of the geographical distribution of labour. The threat of crop failure and famine is ever present in the Kerio Valley. Creating a surplus of grain to be exchanged for livestock appealed to some Marakwet, but far from all. For many, it remained important to complement irrigated agriculture with rain-fed hill farming, bee-keeping, livestock-rearing, foraging, petty trade, handicrafts, contract labour, social alliances, etc.

The geographical areas that Hakansson analysed differ from the Kerio Valley in several respects. Gusii is a high-rainfall area with very productive agriculture. It is densely settled and active markets have evolved along the border with the Luo. The Pare chiefdoms were hierarchical societies with donations of foodstuffs and livestock as integral parts of their economies. Cattle were all-important in marriage exchanges and in building political influence in a way that is hardly paralleled in Marakwet. Pare was, moreover, on one of the nineteenth-century caravan routes. So the situations in western

Kenya and north-eastern Tanzania, on one hand, and in Marakwet, on the other, were rather different, and there is no reason to expect identical causes of agricultural intensification. In Marakwet, irrigation was linked first and foremost to the quest for more reliable subsistence. However, I have also shown that regional exchange, the variable that Hakansson singles out as fuelling agricultural intensification, likewise played a role in the expansion of hill-furrow irrigation in Marakwet. While the processes are not identical, the same factors are operative.

GROWING APART

The geographical division of labour gradually widened the distance between the groups. The current differences in social organization between the pastoral East Pokot and the Marakwet are instructive. While the basic features remain similar – patrilineal descent, decision-making in local neighbourhood councils, egalitarian ideology and mutually intelligible language – there are also notable differences. The risk-aversion strategies characterizing livestock management in semi-arid environments gave rise to egocentric strategies among the Pokot. Relations thus formed are important sources of support in times of need (Bollig 1993, Peristiany 1951). Documenting the coping mechanisms among East Pokot pastoralists after two years of considerable stress, Bollig concluded 'that there is no single type of relation an actor can trust in. He can never tell if his chances to obtain help are better with a patrilineal relative, with an in-law, with a distant affine, with his mother's brother, or with a stock friend. The only thing he may safely assume is that the probability that one or two of his exchange partners might help him is quite high' (1998: 152; see also Bollig 2000). Personal networks and peer groups matter also to the Marakwet of the Kerio Valley, but it is the descent group that is crucial. The 'clan', in its manifold and ambiguous meanings, is what provides land, water and security among the Marakwet.

One notable difference between the neighbouring peoples is that the pastoral Pokot have integrated Karamojong-type initiation into a generation-set system based on circumcision in the Kalenjin tradition, as is practised by the Marakwet, the West Pokot and the Keyo (Beech 1911, Bollig 1990). The ritual specialists are central figures in Pokot society but do not have a similar leading position among the Marakwet, and so on. Very clearly, the different social organizations of these two groups cater for two different ways of life. The contrasting economies have resulted in heterogeneous social and ideological patterning. The economies have diverged, but not to the point of total specialization. Each group has tried to rely on several resources, although specialization has also created resources for exchange. Both strategies have served the need to increase security in an environment in which the risk of drought must always be reckoned with.

The diverging economies have been accompanied by cultural distancing.[18] Thirty years ago, Barth (1969: 16ff.) outlined how ethnic distinctions may emerge in situations of ecological interdependence, making other ethnic groups become, as it were, part of the natural environment. More generally, one constituent dimension of culture is to emphasize differences, to 'singularise', to use Kopytoff's term (1986: 73). Objects of material culture can come to embody ethnic and other identities (see Appadurai 1986, George 1999). The Kerio Valley exemplifies both these processes. Discussing the 'old days', Marakwet elders say that they and their neighbours 'all looked like Maasai'. They emphasize that the different peoples living in the Kerio Valley dressed similarly and that there were few differences between the kinds of ornaments and the artefacts they used. The earliest photographs from the area bear this view out. Today, however, the contrasts are marked, and, as noted above, this is particularly so at a border village like Chesegon. Nowhere else are people as 'typical Marakwet' and as distinct from the Pokot as there.

East of the Kerio Valley, Hodder (1982) identified ethnic boundary-marking in detailed studies of the material culture of the Chamus, the Pokot and the Tugen. Though they all make their stools, gourds and pots, etc from the same materials, their artefacts are strikingly different. Although the material culture was distinct for each people, Hodder also found that many artefact types crossed ethnic borders. The study reminds us from yet another angle of how interaction and exchange characterize the area. Hodder found a 'continual tension between boundary maintenance and boundary disruption; the one exists in relation to the other' (1982: 73).

Conclusion

It has been argued here that irrigation in the Kerio Valley began among small and fluid agglomerations of people with shifting modes of subsistence. In time, a geographically based division of labour between the people who became the Marakwet and their neighbours institutionalized itself and hill-furrow irrigation evolved as part of this process. Movement up the escarpment to the west was restricted, and pastoralists dominated the plains to the east and north. These conditions combined to encourage the expansion of irrigation from the permanent rivers descending the west side of the valley, with now 40 major channels bringing irrigation water to more than 4,000ha of irrigated land. Both the construction of furrows and their management were decentralized and distribution of both land and water was organized by descent groups, guaranteeing access to all their members.

I have attempted to turn the ethnicity argument upside-down, arguing that the ethnic identities developed in the valley as a result of households adopting different subsistence strategies, resulting in a geographical division of labour between the various groups living there. In that situation, it made sense to distinguish between the different groups. While it is true that the colonial administration selected the ethnic labels, which are still in use, and also drew territorial boundaries, it did not invent the constituent groups. They had already evolved.

The geographically based division of labour between the groups inhabiting the valley is the key to explaining how the Marakwet came to construct the spectacular furrows along their escarpment. But to effect this other conditions were equally important. It was essential that descent groups were in command of the distributions both of land and of water, guaranteeing all members access to these resources. This made a labour force available. Another important building-block was the decision-making process. It was decentralized and thereby could directly involve all concerned. Yet it could be multiplied in ways that allowed grand constructions to be achieved. These elements of the Marakwet social organization in particular enabled the people who became the Marakwet to undertake these amazing channel constructions.

The people who settled on the lower slopes of the escarpment became expert furrow-makers. This same process made them different from their neighbours, and their particular social fabric evolved. In that way, the conclusion may be that it was not only the Marakwet who made the furrows but also that the furrows were instrumental in forming the Marakwet.

Notes

1 Much of the field research on which this paper is based was only possible with the collaboration of A.K. Cheptum, curator of the Kabarnet Museum, and of the late J.K. Kassagam, conservator in the Department of Ethnography of the National Museums of Kenya.

We record with sadness the death of our colleague, Johnstone Kibor Kassagam, on 31 March 2003, in his home district of Marakwet, while this book was in press. Besides being a brilliant, inventive and tireless fieldworker, he was instrumental in the development of some of Kenya's regional museums and in the conservation of ethnographic materials from the whole country. It was a privelege to have been able to work with him over a period of almost thirty years.

2 I refer to fieldwork in 1973–5 and subsequent, shorter, research visits in 1976, 1984 and 1998.

3 Wittfogel draws on Beech (1911) to link irrigation and autocratic rule in West Pokot. However, Beech's work is centred on a different economy, pastoralism in East Pokot; it has only three sentences on irrigation (p. 15) and nothing on autocratic rule.

4 See Anderson and Johnson 1988, Berry 1993, Iliffe 1995, McCann 1995, Niemeijer 1996, Richards 1985, Sutton 1989, Waller 1985.

5 Fairhead and Leach 1996, Leach and Mearns 1996, McCann 1999, Schoenbrun 1999.

6 Behnke et al. 1993, Dahlberg 1996, Scoones 1995.

7 See Gupta and Ferguson 1992, Rodman 1992.

8 Ranger 1983 and 1993. An interesting forerunner to that particular discussion, dealing specifically with western Kenya, is Southall's analysis of how the Luyia emerged from a number of small, segmentary groups (1970).

9 As current writings on African history emphasize flexibility and adaptability as characteristic of African societies (see note 4), it is interesting that, when in 1911 the former commissioner for the British East Africa Protectorate, Sir Charles Eliot, discussed the nineteenth-century history of the peoples today designated Kalenjin, he described them as 'not homogeneous … mixed and fluid societies' (in Beech 1911: xii).

10 In the 1930s a government officer wrote that there was 'still far too much forest

destruction, chiefly by the pastoralists, who burn patches of forest in order to cultivate finger millet, which, when harvested, tides them over the dry season when their cows are dry' (Chaundy 1939: 29).

11 Throughout the 1990s and into this century, cattle thefts and attacks on villages have occurred in the Kerio Valley (and parts of the Cherangany Hills). Raiders are armed with automatic weapons and the loss of human life has been severe. Both the clashes and initiatives to put an end to the fighting are regularly documented in *Clashes Up-Date* published by the National Christian Council of Kenya. The fighting during the last decade is exceptional in the very serious consequences it has had and in its continuing such a long time, and also in its direct links to economic and political conditions at a national level (Republic of Kenya 1992, Nowrojee et al.1993, Kagwanja 1998).

12 District Commissioner P.G.P.D. Fullerton, Elgeyo-Marakwet District Annual Report, 1962: 4. (The District Annual Reports are available in the Kenya National Archives.)

13 During a quick and informal survey in 1998 among men of the Korongoro and Sawe generations in Kapchebet village, close to Tot, all of them claimed to have undergone the *sapana* ceremony.

14 Hakansson estimates that, in the Pare mountains in Tanzania, rain-fed cultivation was able to support the population in precolonial times. Yet people still embarked on constructing furrows and developing irrigated agriculture (1995: 308).

15 I. R. Gillespie, Elgeyo District, Handing-over Report, December 1935 (Rhodes House Library, Oxford, MSS Afr.), p. 663. Adams (1996: 160ff..) provides further examples of colonial officers taking the Marakwet to task for 'their lack of enterprise'.

16 The contrast is currently not as sharp as it appears from Gray's book. Nowadays, the water allocation council, at least in some villages, meets every morning of critical times of year to consider the needs of each person's fields (Sutton, pers. comm.). Furthermore, the contrast may be partly an artefact of Gray's training in the 1950s, compared with what is of interest to anthropologists today. (See Adams et al. 1994.)

17 S.F. Moore assumes similar processes with irrigation in Chagga areas and she postulates a 'secondary consequence of ensuring conditions favourable to local trade' (1986: 24).

18 Reid provides an interesting parallel example. On the basis of archaeological, genetic and linguistic evidence he argues that in the Great Lakes region farmers and pastoralists have grown progressively more distinct from each other (1998: 6ff.; see also Reid and Young 2000). A similar reasoning has also been advanced by Schoenbrun (1999). Ehret (2002: 131ff.) finds similar processes at work in the southern parts of the Ethiopian Rift Valley as long ago as 1000 BC. Writing about Maasailand, and closer in time, Waller (1988: 107ff.) states that 'during the nineteenth century, identities in the Rift Valley had been extremely fluid and had depended primarily on mode of subsistence. .... At the end of the century, however, circumstances made a tighter definition of identity necessary and ethnic boundaries became more closely drawn.'

Three

Agricultural Intensification & Social Stratification
Konso in Ethiopia contrasted with Marakwet

ELIZABETH WATSON

Introduction

This chapter presents information from Konso in Ethiopia, in order to revisit the debate about the link between agricultural intensification and social stratification, and to examine its relevance to the region of eastern Africa with which most of this volume is concerned. The impetus for this is twofold. The first derives from the author's work on the social organization of intensive agriculture in Konso, focusing on local priests-cum-chiefs and the role they play in the management of land and labour, and hence in the production and reproduction of the agriculture. Studying Konso from this perspective suggests that social stratification is a feature related to the intensive agriculture. The second impetus is comparative: the Konso situation contrasts distinctly with that of Marakwet in Kenya, which was studied before embarking on the Konso research – and which is, moreover, the subject of Chapter 2 by Östberg in the present volume. Marakwet society is highly egalitarian (relations between men and women apart), and the organization of the agriculture, particularly the administration of the irrigation system, is also markedly egalitarian. Both Marakwet and Konso have been cited as cases of intensive agriculture (Sutton 1989, Amborn 1989, Hakansson 1989) and appear in this volume as 'islands' of the same – which might be taken to imply a degree of comparability, or even similarity. This chapter addresses the question of the extent to which Konso may be interpreted as an example of hierarchical social organization and thence the relationship between this social organization and the intensive agriculture.

However, it is not possible to trace the historical development of this relationship or to outline the evolution of the agricultural system in a way that has been attempted in certain parts of Ethiopia (such as McCann 1995), because of the dearth of sources of evidence and the gross imprecision of any dating pointers in Konso before the late nineteenth century. Oral histories suggest that the people who are Konso today came from all over the region, and the Konso Highlands may have served as a refuge for people from the surrounding areas at times of drought, war or other kinds of social disruption. An archaeological approach, emphasizing environmental and settlement history, might help rectify the rather synchronic view of Konso which currently obtains. There have, of course, been claims in general studies of Africa, employing culture-historical and comparative-linguistic approaches (such as Murdock 1959: 196ff, Ehret 2002: 134, 219), for an antiquity of several millennia for intensive agricultural practices among Eastern Cushitic-speaking peoples in this region. That view awaits archaeological corroboration.

In the meantime, a perspective on the relationship between social organizations and intensification can be obtained by analysing two current examples that are spatially separated. After analysing the Konso material, the chapter returns briefly to the Marakwet case for comparisons, noting differences as well as similarities. The study is based on nine months of fieldwork in Marakwet in 1992 and 1993, and 21 months spent in Konso in 1995 and 1996.

Theories of social intensification and stratification

Theories of agricultural intensification have usually concentrated on the effects of population pressure on land, or on responses to market incentives (as reviewed by Hakansson 1989). They have also examined the relationship between social stratification, political centralization and intensification. In general, discussion of the relationship between social stratification and agricultural intensification has been more concerned with understanding variations in social organization through time and space than with the nature of the intensification in question, that being seen as an accompaniment to political centralization. In some cases intensification has been considered the means through which political centralization occurred, and from which the origin of the state may be traced. This line of thought has a long history, stretching back to Wittfogel's 1957 thesis on the emergence of bureaucracy and despotism.

Commonly such studies exhibit an evolutionary bent, either in a general sense (Service 1975, Sahlins 1974) or in that of an explicit historical materialism (Donham 1990, Friedman 1975, Kahn 1981). Their principal concern has been with processes of transformation and the varying complexity of different societies, and in particular with progression from one stage of production and social organization to another, for example from that of hunter-gatherers or relatively egalitarian extensive cultivators to more complex and stratified societies relying on intensive production (Sahlins 1974). In this outlook big-men and chiefs have been seen as intermediate forms that represent stages in the development of the state (Sahlins 1963, Service 1975).

In reviewing the themes of specialization, exchange and complexity, Brumfiel and Earle (1987: 1ff.) usefully summarize different approaches to understanding the relationship between intensification and stratification. They identify three models – which they label commercial, adaptationist and political – each of which gives a different relative weighting to the economic, environmental and political factors involved in transforming production systems and social organization. In their commercial model, emphasis is placed on the processes and repercussions of economic growth, itself seen as a product of technological innovation. Such innovation and growth result in the production of surplus; in turn, the control and exchange of this surplus lead to social differentiation. By this model, stratification results from the actions of different parties pursuing profit and advantage.

In the adaptationist model, political elites emerge to enable production to be coordinated at a level beyond the household. Such political elites accumulate and redistribute produce across different micro-environments and sectors. This redistribution may occur between highland and lowland areas, for example, or between farmers and craft manufacturers. Frequently, accumulation serves as a risk-aversion strategy, with food or wealth being accumulated in good years and redistributed in those with poor harvests. In such situations political elites 'enhance productivity, diversify subsistence and provide insurance against food failures in one sector or another' (Brumfiel and Earle 1987: 2). But their role extends further, notably to facilitating centralized decision-making. This enables land and labour to be scheduled for maximum use, and also permits agricultural works which require the co-option of many people, for example the construction and maintenance of irrigation canals or terraces. In this model, stratification is viewed as serving the interests of the many, and enabling improved use of available resources. This is in contrast to the third, political, model, in which stratification and intensification result from individuals' quest for power. Here 'political elites consciously and strategically employ specialization and exchange to create and maintain social inequality, strengthen political coalitions, and fund new institutions of control' (ibid.: 1987: 3).

The drawback of these models is that they can be used to imply a determinism where the form of society is driven by a desire for profit, for more effective production or for power. This is not the aim here; the aim is to appreciate the connections between political stratification and intensification and why or how they arise together on the ground. The emergence of political stratification is likely to be connected to various economic, practical and political factors, in the sense that Donham proposes 'the positive consequences of Y for X are part of the explanation for Y' (1990: 72), but they are only part of the explanation, and X is only one possible outcome. (Donham also discusses the difference between functional explanations and functionalism.) The models discussed above are useful to this study because they make explicit some of the factors that may contribute to stratification, and it is likely that stratification is connected to a combination of some or all of these different factors, rather than only one or another.

Intensive agriculture in Konso

The Konso occupy a small range of hills which rise to a height of 2,000m, some 600km south of Addis Ababa (Figure 3.1). The population is currently estimated at some 215,000. Rainfall in these hills is rather low and unreliable. From 1977 to 1994 the mean annual rainfall was 551mm, but this varied between extremes of 282mm and 881mm. Moreover, the rain is apt to fall unevenly, or 'at the wrong time' for farmers. As a result, the Konso have been frequently affected by food shortages (Messeret 1990). In 2000 Konso was one of the areas of the country worst hit by drought.

The Konso employ a variety of soil- and water-conservation techniques to facilitate their cultivation of the hillsides and to maximize the available sources of water. These include mulching, constructing square-ridged basins, and other techniques which are delineated below. The most immediately obvious of these, and most striking to a visitor, are the drystone-walled terraces which cover almost the whole of Konso (Figure 3.2). The resulting landscape is impressive, especially in contrast to the surrounding dry lowland areas, so that Konso appears like an island of intensive agriculture. The Konso themselves are proud of their landscape and their agricultural practices. Generally speaking, they value the work of digging and constructing terraces above all other activities, and the importance that they place on stone structures has led several commentators to describe them as 'megalithic' (Jensen 1960, Amborn 1989, Straube 1963, Ministry of Agriculture, Ethiopia 1988).

It is difficult to determine the age of this form of agriculture, a question which has vexed scholars who have worked in the region. Amborn, in his broader study of the Burji-Konso cluster, estimated that Konso agriculture is at least 400 years old, basing his deduction on genealogies, constructed by himself and others (1989: 73).

The main features of this agricultural system may now be summarized (following Amborn 1989, which can be consulted for detail and references to earlier studies):

Terracing. Drystone bench terraces (*kawwata*) are the most important soil- and water-conservation structures (Figures 3.2 and 3.3). These are built on land at around 1,000m and above, which in practice means most of Konso core territory. The sizes of terrace walls depend on the terrain and, though heights of 0.5–1.5m are preferred, field walls up to 6m high can be observed in steep places (Amborn, 1989: 74). The terraces prevent soil erosion and maximize water retention and infiltration. Sometimes they have drainage outlets, to prevent damage from stormwater and to improve irrigation to lower terraces.

Irrigation and rainwater harvesting. Small ridges, furrows and terrace walls are constructed to channel rainwater from roads, paths and other areas on to the fields. On the sides of temporary or minor streams, small canals with stone and mud banks are built to carry water when the stream fills. Stone dams are

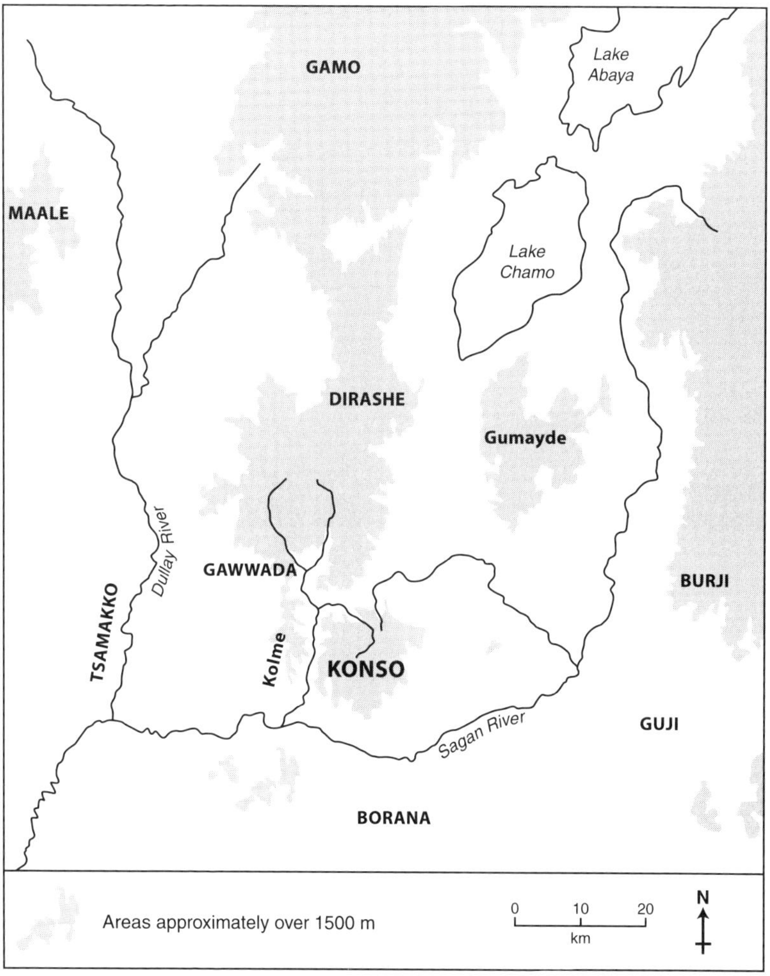

Figure 3.1: *Map showing Konso in south-western Ethiopia and location of neighbouring people (in capital letters). Gumayde and Kolme are areas where many Konso have recently settled. (Map: Department of Geography, University of Cambridge).*

Figure 3.2 *Konso terraced hillsides (All photographs: Elizabeth Watson)*

Figure 3.3 *Drystone cultivation terraces, with square basins and drainage channel*

occasionally built across such streams to raise the level and enable water to be led to the fields. During heavy rain, these small dams can be quickly washed away. There are few perennial streams, and irrigation mainly consists of such opportunistic rainwater harvesting.

Soil bunds and square-ridged basins. Earthen ridges, 10–20cm high, may be constructed across the terraces, thus dividing them into small square-ridged basins. The ridges are covered with sorghum stalks and other trash. These square-ridged basins (Figure 3.3) are very effective in maximizing infiltration.

Zero-grazing, stall-feeding and manuring. Cattle are stall-fed in the homesteads in the villages (Figure 3.4), the latter comprising distinctive, densely packed, walled settlements. Manure is collected from the animal pens, which are built on the lower sides of the homesteads, and taken out of the village to be spread on the fields (Figure 3.5). Since most of the fields, especially those around the villages, are rarely left to fallow, this manuring is essential for maintaining fertility. In some villages dried human faeces were formerly collected and applied to the fields as fertilizer, but this practice is now rare (and serves as a source of ridicule about other villages).

Intercropping of diverse crops. The crops are mixed in the fields, and they are also combined with trees and bushes, such as *moringa* (whose leaves are eaten as a green vegetable), coffee and *qat* (which is grown for sale). The Konso rely heavily on drought-resistant crops, the principal being sorghum, but there is a remarkable variety of other crops, such as finger-millet, maize, wheat, barley, sunflower, castor, beans, pumpkin, tomato, chilli, cotton, potatoes (both sweet and Irish), cassava, bananas, tobacco, gourds and garlic. Most of these are grown throughout the Konso Hills, but there is some variation with altitude (as detailed in Westphal 1975). Wheat and barley are not grown below 1,800m.

The combination of these diverse techniques maximizes the use of available water and land, but also minimizes risk in what is a high-risk environment. This diversification can also be seen in the broader approach to livelihood strategies. Although the main activity is farming, this is supplemented by other activities. Some people spin and sell cotton, brew and sell beer (women only), trade or, occasionally, hunt. There is also a minority artisanal group known as *xawuda*, who in the past did not farm, but made their living as potters, blacksmiths, tanners, butchers and weavers (Hallpike 1968, 1972; Watson and Lakew 2001). These craftspeople formed a separate social category which was culturally debarred from intermarrying with the main farming population. Now the distinctions are breaking down, so that artisans and farmers are, to an extent, integrating and diversifying into each others' occupations.

Figure 3.4 *Cow in stall in homestead*

Social organization of labour

'Intensification has a cost ... the cost is often in human energy, the product of sinew and sweat.' (Netting 1993: 103).

The construction of these soil- and water-conservation structures places obvious demands on labour. Measuring the net extra labour costs of intensification and of the construction of landesque capital has been shown to be extremely complicated and difficult (see Netting 1993, Blaikie and Brookfield 1987). Despite this, it is obvious, intuitively, that a field that relies on such intensive techniques needs the investment of more labour than one that does not.

Each Konso household is responsible for gaining access to sufficient labour to build soil- and water-conservation structures on its fields. Since it is the individual household that benefits from the quality of the fields, investment in these structures is rarely a wider community concern. There are no sanctions to ensure that a household adequately maintains the structures, and if a farmer suffers because of the inadequate maintenance of terraces in a neighbour's field, there is no means of enforcing the desired action.

Men, women and children work together in the fields, and a number of tasks can be performed by men or women. (Children generally perform tasks associated with the adult of their sex: in this way they learn the techniques and also make an important labour contribution.) Both men and women dig, sow crops, weed, harvest and scare birds. Carrying and spreading manure is a task usually associated with women (Figure 3.5), but there is no prohibition

Figure 3.5 *Odiema Xalale, a girl from Buso village, distributing manure on a field*

on men performing it, and they often do. Terrace construction is mostly the work of men, but if a large amount of work has to be done, men and women may work together at it. In such cases, men do the actual stonewall building of the terraces and pack the necessary earth around them, while women use wooden bowls (known as *toma*) to move earth to level the terraces, and to collect stones which they pass to the men. It is thus a cooperative process. Women's labour is also very important when a work party is called to a field. The workers must be provided with something to eat – or at the least *chagga*, the Konso grain beer famous for its strength-giving properties, and sometimes beans too.

Brewing beer and cooking food is exclusively the work of women. Beer takes around three days to be brewed and consumes a lot of labour, because sometimes as much as 50 litres must be prepared. Without beer there will be poor attendance at a work party, and the work itself is likely to be lacklustre. These work parties sometimes consist exclusively of women, but only rarely exclusively of men. Most supposedly male work parties usually have at least two women whose job is to bring water to the workers in the fields, and to carry the lunch of *chagga* and beans if necessary. Thus most of the visible work of constructing the terraces and other forms of landesque capital tends to be performed by men, but the process includes a great deal of equally important though less visible work which is undertaken by women.

Each household works on its own fields, but if possible it will call on extra labour through work parties. The amount of labour a household can recruit in this way depends on its own participation in others' work parties, as well as on its power and influence. It is through a closer examination of the work parties that the relationship between intensive agriculture and social differentiation can be appreciated.

Konso work parties

Amborn has described generally how individuals gain access to labour along kin, generation-grade and residential lines (1989: 80). However, the nature and form of the specific kinds of work parties deserve examination in more detail, in order to illustrate the social relations through which Konso intensive agriculture is produced. Five different forms of work parties have their own names:

Koda fadeta is the most common and important form of work group. It is comparable with the festive work group of the Maale people living some distance to the north-west (as described by Donham 1985). In Konso, *fadeta* implies assistance or pulling together. It is informal and involves calling friends, relatives and neighbours to assist in work. No one is compelled to attend when a person calls *fadeta*, but a person who receives assistance is tacitly obliged to return the favour at some time in the future. The workers will come to the caller's house before and after the work for grain beer, and the caller must send grain beer to the fields for lunch. If a person is liked and has good connections, then perhaps 20–50 people will go to the fields for him or her (Figure 3.6).

Fadeta is called as a matter of course during the farming year, but is also called if a household experiences difficulties in finishing its work because of misfortune, for example through sickness or bereavement. (In the case of the latter, the family is obliged by custom to observe a formal mourning period during which work is forbidden, even if it occurs at a critical time in the farming calendar with the danger of the harvest being ruined.) This illustrates the important, if publicly unrecognized, reciprocal nature of this form of work. People participate in *koda fadeta* when they are called by their friends, relatives or neighbours, because survival may depend on such mutual assistance. Conversely, a person who consistently fails to attend *koda fadeta* will find no one willing to help when he or she needs it, whether for everyday work or in special circumstances. Though it is couched in terms of assistance, *koda fadeta* involves a complex web of ties of mutual obligation.

Koda kaffa literally means lineage work; such work parties are formed when a person calls his lineage or clan to assist. It is used in cases of severe difficulty and is not particularly common, as it may involve the mobilization of a large number of people who must be provided with beer at lunchtime and in the evening.

Koda kanta literally means neighbourhood work. If a neighbourhood agrees to help a person by working for him/her, then all members of the neighbourhood must attend the work-day; if a person fails to attend he/she will be fined. Like *koda kaffa*, this is not as common as *fadeta*, because the neighbourhoods have large populations, usually of 110–150 households, and

Figure 3.6 Koda fadeta *(work party) preparing terraces and ground for planting*

Figure 3.7 *Flood-protection wall* (tomota)

the beer for so many people is expensive. It is a residence-based form of work party, but the person requesting the labour may not necessarily be a resident of the neighbourhood called. Essentially, a household contracts a whole neighbourhood to work for them in return for beer. It appears a point of honour on the part of the neighbourhood to ensure a good turn-out for this kind of labour and to do the job well. During the writer's stay in Konso in 1996, one household called two neighbourhoods to work for them on two different occasions. This involved brewing four vats of beer on each occasion (costing approximately 200 Ethiopian *birr*, the equivalent of £20 sterling, each time). A labour force of this size was needed to reconstruct some massive flood-protection walls that had been destroyed by heavy rains (Figure 3.7).

The remaining categories of work groups are cyclical and quite common. They are known as *parka*, for a small group, and the larger *marpara* (the latter term derived from the Amharic *mahper*, meaning association). During any particular year these groups have a fixed membership, the members working cyclically on each other's fields and also on non-members' fields for money. The money earned is pooled and used to buy an ox at the end of the farming season. This is slaughtered with some ceremony, and then eaten communally. The size of these groups can vary from around eight people for a *parka* to as many as 50 for a *marpara*.

There is, as shown, some degree of variation in the formality of the membership of the different kinds of work parties. Whereas *koda kaffa*, *koda kanta* and *marpara* and *parka* are quite formally organized with fixed membership, *fadeta* is more informal and negotiable. Membership of *koda kaffa* and *koda kanta* is based on kinship and residence respectively. These factors tend to cut across social differences, including that of wealth. Membership of *marpara* and *parka* tends to be based on friendship, but commonly members are from the same generation-grade. Membership of these groups may also vary over time, as friendships form and disintegrate.

The *koda kaffa*, *koda kanta*, *marpara* and *parka* groups embody essentially an egalitarian principle, as each person contributes equally to these work parties, and there is no discrimination on attendance and participation. In all these groups the participants are of diverse social status. This egalitarian tendency, in conception as well as execution, is borne out in Amborn's study. He found that the working parties discussed above, based on kin, neighbourhood and generation-grade, cross-cut each other in the Burji-Konso cluster so that 'the structure of the political system is characterized by factors designed to counteract tendencies of individual groups to gain dominance. The kinship system, the territorial units, the generation-group system and, with it, all offices of political and religious affairs interact to balance one another' (1989: 81). However, that analysis was based on observations of the composition and participation of work parties. Further examination of the functions of the work parties – in particular of who uses them and who benefits – shows that this egalitarian picture, while essentially valid, may not reflect the full story. This prompts us to look in more detail at *koda fadeta*.

For some people, participation in extra-household labour parties is cyclical, so that there is no net individual gain. Such access to extra-household labour is valued more for social reasons and efficiency, and can enable a farmer to construct or maintain terraces on his or her fields over a matter of days, something which might have taken the household weeks or months if working alone. In general, the labour of moving stones and soil is found to be easier if undertaken by a group, with the chatting and mutual encouragement thus generated, than it would be in isolation. Other people are likely to experience a net loss of labour, if anything, whenever they participate on others' fields in *koda kanta* or *koda kaffa*. In the cases of *marpara* and *parka*, members of the work parties benefit from a free turn whenever the group works on their own fields, but they also work for others for money (some of whom may actually be members of the groups, and who get a 'members' discount' on a day's labour). Some people lack the resources to brew beer and are thus unable to call *koda kaffa* or *koda kanta*, or to hire a paid work group, whereas those who can afford this are able to benefit from a net gain in labour applied to their fields.

Thus the extra-household labour executed by work parties must be understood in the context of the wider economy and of the various exchanges that occur between different people within the community. By examining the situation in this way, it transpires that there is considerable variation in the positions of individuals, the exchange of labour and other products being exploited differentially.

This can be demonstrated in better detail by a study of those individuals known as *poqalla*. These are priests-cum-chiefs who fulfil an important role in Konso society, their differential social status being institutionalized. Their participation and use of work parties, particularly that of *fadeta*, casts Konso intensive agriculture in a perspective rather different from the egalitarian principle noted above.

Poqalla: Konso priests-cum-chiefs

Poqallas make up approximately 3 per cent of the population of Konso, but their level of importance varies widely. There is, nevertheless, a consistent logic and form in their social position and functions at all levels (for detail, see Shako 1993 and Watson 1997). *Poqallas* are hereditary male leaders who are lineage or clan-heads as well as priests. Their religious, ritual and priestly role is emphasized in Hallpike's ethnography (1972). But in fact their social, political and economic influence is equally important. *Poqallas* participate actively in the labour exchange described above, and often benefit from a substantial net gain of labour. An examination of how labour is managed reveals their role in agriculture and in the production of Konso's impressive landscape.

Poqallas are credited with being the first-born descendants of the founders of their clan or lineage. All people in Konso are therefore considered to have descended from one *poqalla* or another, whom they regard as 'their' *poqalla* and should respect as such.

As priests, they are mediators between *Waqa* (God) and the people, performing ceremonies and blessing the people (Figure 3.8). Through such ceremonies the *poqallas* are seen as keeping the culture, thus ensuring that people, crops and livestock prosper. In this way *poqallas* embody and are able to control the fertility, defined in the broadest sense as wellbeing and the productivity, or reproductivity, of their clan or lineage.

As keepers of the peace, *poqallas* also have a political role. They mediate in disputes between individuals and can fine perceived wrong-doers. They can mediate in wars and battles, either by themselves walking between the protagonists, or by sending a representative to bring peace. Conflict and quarrelling are thought to displease *Waqa*, and thus result in rain being withheld and the concomitant drought, hunger, infertility and disease.

One aspect of the *poqalla*, which has been previously overlooked, is their control of large amounts of land. In the main, land in Konso is individually held, and can be sold or bought, inherited, lent or given away, virtually as private property. There are, however, obligations connected to owning large amounts of land, and here the *poqalla* comes to be seen as the caretaker of the property of his lineage. On balance, however, he seems to have the right as an individual to dispose of the land as he wishes, as if by exclusive ownership. In this sense *poqallas* can be considered large landowners (a subject developed in Watson 1998).

The *poqallas* farm much of the land they hold themselves, and lend land to others who ask for it. Those who borrow land are generally the needy, often younger brothers with little land of their own because inheritance is unevenly spread between male children. Widows and single women may also borrow, as women in Konso cannot inherit land and generally do not own it. They are therefore very vulnerable; a widow becomes dependent on her husband's family continuing to support her, or on her father. In some cases, a woman may not receive land from either side, in which case she may ask a *poqalla*.

Control over these resources enables *poqallas* to assist people with difficulties, especially shortage of land to cultivate; similarly during famine individuals may request grain from a *poqalla*. The *poqalla* can also help with less frequent demands that arise during the course of an individual's life: at marriage he may give the cowhide or butter that a new couple need for the marriage ceremony and for their future life; or he may help a younger brother with stones (obtained from his many fields) to build a house, or with grass to thatch it. Parallels for this moral obligation of landlords to provide succour for poorer people and assistance in times of dire need can be found in other countries (see for instance Scott 1976, on the 'moral economy of the peasant' in south-east Asia). In the Konso environment, with its vulnerability to crop failure, this method of providing food relief and other assistance is invaluable to poorer families.

Thus the different roles of the *poqalla* – priest, political leader, clan-head and landowner – are intricately related. Their religious, political and economic powers are not easily distinguished. Flows of prestations go from the *poqallas*

Figure 3.8 *A* poqalla *beside a granary, waiting to receive people who will go hunting as part of the* kara *ritual. The wooden bowl contains milk and a bundle of auspicious plants* (hansabita), *which will be used in blessing*

to other members of the community. Blessings flow from the *poqalla* to his clan, which ensure fertility and wellbeing. Actions and rulings favouring particular people amount to a form of patronage. Land may be given, as well as food and other goods, at times of hunger or other need. (See Watson 1998 and, for parallels among Maale kings and chiefs, Donham 1985.)

In return for such gifts or prestations, the *poqallas* receive labour. This may be prescribed and direct, or more informal. In a case of land lent to a household, the labour to be provided by the borrowing household is well-defined, occurring three times a year, at sowing, weeding and harvest. In the past this was very strict and known as *piyolada*. It was outlawed after the Ethiopian revolution of 1974. But it is still practised, though less formally and prescriptively than before, and with the name *piyolada* having been dropped (Watson 1998). There is no direct labour prescribed for other gifts, such as grain or grass or butter; but a *poqalla* can call his lineage or clan division to labour for him, and many people willingly heed the call, saying that they are honouring him. Although this could be interpreted as provision of labour in return for blessings, it is equally an illustration of how the clan associates the prosperity of its members with the *poqalla*; they will work for him to help him prosper, even though there is no compulsion or specified labour obligation. However, a *poqalla* will not necessarily assist anyone as a matter of course: it is said that he will help those with whom he has good relations, and a person can make relations with the *poqalla* by going to work on his fields when he calls. In effect, then, labour is provided in return for gifts. Such labour is called *fadeta* (as defined above), and most *poqallas* possess the resources to provide their work parties with the necessary beer.

The *poqalla* thus relates to his lineage or clan through these flows of exchanges, blessings and goods for labour; but he may extend his power and influence further by acts of generosity, such as lending land or goods to people outside his immediate circle of descendants. *Poqallas* may also be involved in political decisions involving members of other clans, and this is another way of asserting influence, and again of obtaining labour in return for patronage. Thus, following Sahlins' characterization (1963) of different kinds of cultural leaders, the *poqalla* acts in some ways like a chief, but in others like a big-man. He fits the first category through his position in a network of exchanges, in particular his receipt of tribute labour from his lineage or clan, which is prescribed because of his birth and position. But he may also fit the category of big-man by gaining followers from outside through political skill and conspicuous generosity, so that people who are not prescribed by birth to give labour or to support the *poqalla* become drawn into his network and augment his available labour force.

The wealth of individual *poqallas*, and the extent to which they fulfil these roles in practice, vary greatly. Since the 1974 revolution, many have lost land and no longer perform rituals or act as political leaders (Watson 1997, 2002). Despite this, many still have access to significantly larger amounts of extra-household labour than do ordinary Konso. This was borne out by a sample survey undertaken in 1996, in which the *poqallas* studied obtained a mean 283 extra-household person-days of labour during the year, compared with an average of 31 days for non-*poqallas*. Equally interesting were the contrasting ways in which this extra-household labour was employed. Whereas commoners used the bulk of the assistance for essential field tasks, such as sowing and especially weeding, which become critical at times in the agricultural calendar, *poqallas* were using a much higher proportion on terracing and other soil- and water-conservation structures. Thus, their access to significantly higher amounts of labour allows them to invest in constructions and improvements to the land, rather than concentrate on cropping alone. In this way the *poqallas* do more than merely maintain existing landesque capital; their investment in labour actually creates new capital.

Intensification and Social stratification: the Konso case

Poqallas are clearly advantageously positioned in the network of exchanges. As seen, they have access to more labour, which in turn enables them to cultivate land and harvest more surplus, which they can then reinvest in attracting more people to work for them as *fadeta*. There is thus an element of social stratification in Konso which contributes to the construction of this intensive form of agriculture and the production of the terraced landscape.

Returning to the discussion of social stratification and intensification,

Konso agriculture and the role of *poqallas* can be accommodated within the models noted (of Brumfiel and Earle 1987). The techniques of cultivation and local scientific knowledge about soil- and water-conservation, as well as the maximization of yields, all contribute to the production of a surplus which is controlled by the *poqallas* and exchanged. This picture fits the adaptationist model, with the *poqallas* organizing production and controlling a surplus beyond the level of the household. They also constitute an institution which helps reduce vulnerability to hazard. Further, power struggles take place as various important people and *poqallas* compete for followers, that is for both political supporters (in the broadest sense) and labour. In this way the political model also has some application here.

In fact, the practical demands of the agriculture, the desire of powerful individuals to increase the amount of land they cultivate and the political need for followers intersect. Land is a limiting factor, of course, but not the main one. Powerful individuals can gain access to land if they wish, and can even own extra land and hold it in reserve. Access to labour emerges as the most important factor in production, as it is needed for investment in the soil- and water-conservation structures that maximize the harvest. The networks of exchanges are used to gain access to this labour. This process is played out in typical big-man and chiefly politics.

While stressing here that the *poqallas*, who were in previous studies described as essentially priests or clan-leaders, play a significant role in the construction of this landscape, it is not being suggested that they are wholly responsible or that they have absolute control over land and labour. As noted, there are alternative ways of organizing work groups, and many individuals farm without recourse to a *poqalla*. There are, moreover, other channels of religious, economic and political power present in Konso. This has become particularly evident in recent years with the growth of Christian movements, the increasing influence of the state and local bureaucracy, and the government's land reform policies. As a hierarchical and hereditary institution (based on primogeniture), *poqalla*-hood remains influential, but is not – and seems never to have been – pervasive in all aspects of Konso life. In fact, some men of the *poqalla* class are quite weak and poor; power depends on a combination of resources and political skill, and can vary markedly, even between generations of the same line.

Comparing Marakwet and Konso

The purpose of this brief final section is to examine the question, noted at the beginning, of why agricultural intensification involves social stratification in some instances, as observed in Konso, but not in others. Marakwet in Kenya provides a striking contrast – as explained in Östberg's chapter (see also Soper 1983, Ssennyonga 1983, Adams et al. 1997, Watson et al. 1998). The Marakwet combine rain-fed cultivation of drought-resistant crops with intensive irrigated agriculture. The irrigation relies on perennial and seasonal streams which descend the escarpment from the high Cherangany hills to the Kerio Valley. Water is redirected from these streams along furrows constructed from local materials (mainly stones, earth and wood). These furrows remain the property of the descendants of their original constructors, which in practice makes them the common property of the various villages, clans or lineages. These owners not only use the furrowed water for irrigating their fields, but also carry out all the necessary operating and maintenance. If these require decisions, then a meeting takes place of all the circumcised men forming the unit that owns the furrow; the decisions are reached through debate. The system is distinctly egalitarian, an effective local form of democracy.

The essential difference between Marakwet and Konso is in the ownership of the landesque capital, that is the infrastructure – notably the irrigation furrows in the one case, and the terrace walls and other conservation devices in the other – on which the intensive agriculture depends and whose construction requires communal or extra-household labour. In Konso, as shown in this chapter, much of the labour which is organized at an extra-household level is invested in the soil- and water-conservation structures on fields belonging to individuals or individual households. Those with more land and wealth (and a modicum of political skill) can turn the 'labour exchange' to further advantage by using it to improve their land. Such arrangements tend to promote social differentiation.

In Marakwet, however, labour organized at an extra-household level is invested in the water furrows, which belong equally to all who work on them and use them (Adams et al. 1997, Watson et al. 1998). Labour is therefore invested in publicly-owned structures which are used to the common good. A whole village or clan benefits from the construction and maintenance of the furrow; everyone needs to cooperate for the communal wellbeing. Labour is important, but is controlled by the community, not by any individual.

While in general one may discern a relationship between the evolution of stratification and increasing privatization, the nature of the resources strongly influences the way the relationship develops. Increasing privatization means increasing exclusive use or, in other words, increasing ability to exclude others from using that resource. Terraced land lends itself relatively easily to exclusive use: the labour invested in building terraces is likely to be translated into increasingly exclusive rights to that individual patch of land. With neighbouring households exercising similarly exclusive rights to their patches, a landholding class interest emerges. Such development of exclusive rights need not automatically divide society into groups of owners and indigents; at least, to be sustainable, it has to function in a way that makes provision for all without the danger of undue conflict erupting. In the case of rights and access to water, by contrast, the social implications of private or exclusive ownership would be much more divisive and serious, in particular if a small group of people were to own a furrow and its water, with the power to exclude others whose fields were dependent on that furrow. Such a form of control would need a much stronger power structure than exists in Konso. There, the institutionalized stratification, represented by the *poqallas*, is carefully balanced by more egalitarian organizational features.

Arguably, in Marakwet, the factors are lacking which might generate the level of power and control needed to sustain exclusive furrow ownership with social stratification. Not only are opportunities of exploiting the regional market poor, but more generally local conditions enable individual households to be more independent. In particular, Marakwet have developed a granary type which enables finger-millet to be stored up to nine years, so that households or kin groups take more responsibility for handling years of scarcity than in Konso, with less resort to succour from richer and powerful neighbours. In Marakwet, the livelihood options are more diverse than those available in Konso: the irrigation, the fertile soils and the possibility of cultivating and irrigating crops in a variety of agro-ecological zones conspire to give the individual household more options and control of its own resources than in Konso. Marakwet do not need to rely on risk-averting supra-community institutions like chiefs and big-men in the Konso manner, and therefore they do not invest in them in the same way.

Clearly, these East African islands of intensive agriculture do not conform to a single model, but have developed highly variable forms, features and social organizations. The differences must result from combinations of factors, both historical and persisting, only some of which have been examined here; population density and a concomitant pressure on resources must also obviously be relevant. Moreover, the relationship between intensification and stratification, as examined in this contribution, emphasizes the importance of identifying the particular resource parameters within which the social organization of production operates. The nature of the resources in question makes a difference to both the kind of intensification and the social organization.

Four

The History of Iraqw Intensive Agriculture Tanzania

LOWE BÖRJESON*

The limits of congested space: confronting theory and history in the Iraqw homeland

The historical geography of the agro-pastoral Iraqw in north-central Tanzania revolves around their highland settlement on the Mbulu Plateau, just above the Rift wall, in Mbulu District. Precolonial Iraqw settlement is, specifically, associated with a relatively small area, about 180 square km, which has been variously described as the Iraqw heartland, homeland or cradle area.

Even now, the agricultural practices and landscape in this historical core area differ from those of its surroundings. Inside, labour-intensive hand-hoe cultivation, including wet- and dry-season cropping, erosion prevention, water control, manuring, composting and mulching, together with cattle-rearing and tree-planting, are the main activities shaping the landscape. Outside, the landscape is generally more gently undulating, the soils are more fertile, but precipitation is less, farming is more capital-intensive and market-oriented, ploughing by oxen or tractor is widespread, and other capital inputs such as chemical fertilizers and pesticides are utilized to a greater extent. Consequently, Iraqw farm holdings outside the homeland are usually larger, and social stratification is more pronounced there than inside it (Thomas 1977: 22-7, Meindertsma and Kessler 1997: 35-68, Loiske 1995: 59-81, Raikes 1975: 124-62, 228, 257-86, Lawi 1992, Lindström 1988: 15-77). In the nineteenth century, the Iraqw population was mainly concentrated in and around the homeland, and the rest of the Mbulu Highlands was dominated by pastoralists and more sparsely populated. However, these differences in agricultural production cannot be explained simply as adaptations to environmental

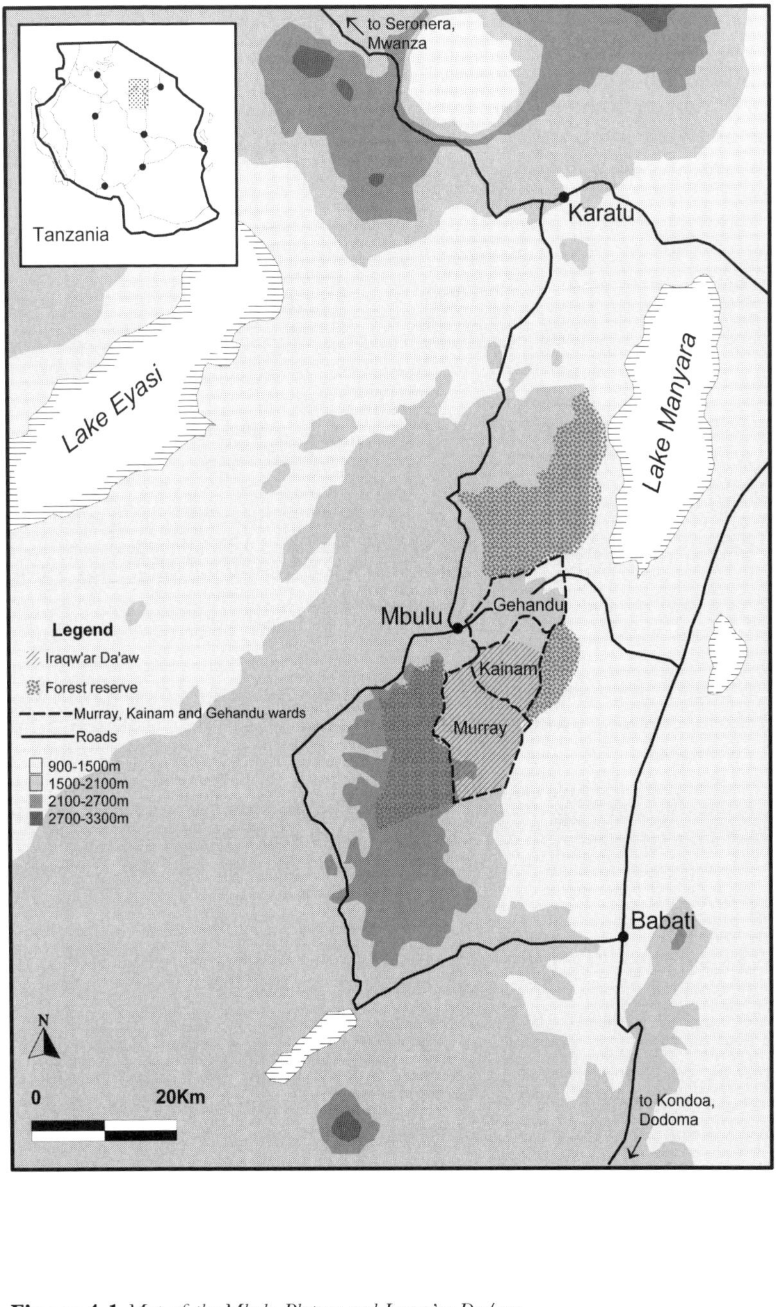

Figure 4.1 *Map of the Mbulu Plateau and Iraqw'ar Da/aw*
(Map: K. Svae, based on Meindertsma and Kessler 1997: 2, 98)

conditions, population pressure and market demand. The background is much more complex: both market development and the production of agricultural landscapes have a political and social history.

The identification of an Iraqw homeland can be traced to a massive expansion of Iraqw society, during the colonial and postcolonial era, giving rise to a division of Iraqw territory into homeland and expansion areas. But, as the discussion below indicates, the notion of a homeland is mainly related to twentieth-century cultural and political representations of Iraqw history, and is thus of little value for understanding nineteenth-century Iraqw settlement dynamics.

Around 1900 Iraqw settlement in the Mbulu Highlands lay broadly to the south and east of where Mbulu Town now stands (Figure 4.1). Earlier in the nineteenth century it had probably been more concentrated, but located within the same area. Datoga and Maasai pastoralists dominated the rest of the Mbulu Plateau during the nineteenth century, and the Iraqw settlement was only a minor enclave of agro-pastoralists bordering the lands controlled by these two pastoral societies. This political geography is reflected in Iraqw oral tradition and history as well as colonial historiography (Bagshawe c 1920: 35-8, Sturdy 1936, Hartley 1938, Schultz 1971: 90-100, Berger and Kiessling 1998: 146-87, Lawi 1999: 34, 65-6). In both colonial and Iraqw source material there is an emphasis on the threat and pressure exercised on Iraqw society by its pastoral neighbours, which in turn has led to the conclusion that coercive force was an effective barrier for Iraqw expansion. Faced with this barrier, the Iraqw, by this view, were imagined as having been enclosed during the nineteenth century, and to have lacked the opportunity to expand until the power of the pastoralists was weakened and their territory contracted at the end of that century. However, research among the Iraqw and the message of comparative studies demonstrate that East African precolonial societies depended on connectedness and versatility, and that hostilities between groups of people were usually intermittent and only partially disruptive. (See for example Koponen 1988a, 1988b, Kjekshus 1996, Maddox et al. 1996, Berntsen 1976, Lawi 1999, Thornton 1980.)

In assessing the role of coercive external force in Iraqw history, one must also consider the gender bias of the primary historical sources. Iraqw oral history and traditions are essentially the products of men, and colonial accounts of Iraqw history are entirely written by men. And, since political and public affairs, as well as historical interpretations, have been dominated by men, the emphasis on coercive political relations in Iraqw history must be analysed in this context. For example, the protection of land, livestock and people from external threat, by means of both ritual power and military defence, is mainly associated with male activities. Men have also been responsible for the governance of natural resource utilization, thus holding much of the formal economic and political power in Iraqw society (Lawi 1999: 187, Thornton 1980: 85-103, Snyder 1999). However, there is also evidence of women's active involvement in political affairs in Iraqw society, revealing strong but less formal means of exercising power (Snyder 1997,

1999, Loiske n.d.). But it is not only the source material that has a gender bias; the study of labour-intensive forms of agriculture has itself been problematic in this respect. Not only have explanations related to political organization, power structures and military matters emphasized the male contribution to intensification, but also the frequent vivid descriptions of physical hard work and monumental investments in landesque capital echo a male bias in the research focus. Undeniably, this bias has consistently overshadowed the more incremental and nitty-gritty everyday work contributing towards intensified land use, by women, children and men alike.

More specifically, however, our concern here is the evolution of intensive farming practices in the Mbulu Highlands in relation to Iraqw settlement history. The Mbulu Plateau is not unique among mountainous areas in East Africa in having experienced precolonial intensive agriculture and settlement concentration (Turner et al. 1993: 27-30). In the Iraqw case the intensive agriculture has been explained since the early twentieth century as a consequence of high population density. Basically, the line of argument has followed the logic contained in the phrase 'necessity is the mother of invention' (Mbulu District Book, e.g. notes on Botany and Agriculture, Sturdy 1936: 53, Hartley 1938, Meek 1953: 159-60, Schultz 1971: 93). Thus, in its most simple form this population pressure theory contends that when land for food production becomes scarce, as population increases without any possibility of settlement expansion, people develop more intensive farming practices rather than starve (see Boserup 1965, Netting 1993, Turner et al. 1993). Therefore, in order to demonstrate the validity of this theory in the Iraqw homeland, as well as other pockets of historically high population densities, possibilities for expansion must be shown to be limited. In the Iraqw case, as portrayed in earlier literature, this criterion has been satisfied by the presence of Maasai and Datoga pastoralists, who have been seen as keeping the Iraqw more or less under siege during the nineteenth century. In fact, Iraqw intensive agriculture is often cited as a typical example of a precolonial siege situation.[1] Southall (1961: 161-3) notes other people inhabiting mountainous areas in East Africa that have been similarly considered as having been under siege by the Maasai during the nineteenth century, following the latter's expansion: 'There were on the mountains large tribal islands washed but not engulfed by the Masai sea: Arusha and Meru, Chagga, Pare and Sambaa, Kikuyu, Meru, Embu, Kamba and Teita' (p. 161).

Indeed, comparable examples of people under siege have been discussed in other parts of Africa and around the world. The settlement history of the Kofyar in Nigeria, for instance, has been described in much the same way as that of the Iraqw (Stone 1996: 59-60).

In order to understand intensification, we need to observe the process and not merely the level of intensity. And as many researchers have discovered, this is not easy even in a long time-perspective (Morrison 1994, 1996, Leach 1999). An observed correlation between areas with high densities of rural populations and high levels of agricultural intensity is not necessarily an indication that population pressure drives intensification. In addition we need to know to what extent the intensive agricultural practices, or rather the socio-economic, political and environmental circumstances under which they have emerged, attract high densities of populations (see Chapter 2). The question is whether the line of causation in population pressure theory may in fact work backwards. Agricultural intensification would in this respect be regarded as an attractive option, with the potential of increasing livelihood security and living standards, and thus in turn of stimulating population growth, including in-migration (see Morrison 1994: 116-36). In many cases, moreover, it is difficult to draw the line between the two central concepts of intensification and expansion. In theory the spatial component in the analysis needs to be fixed, but in practice this is almost never the case. More commonly people tend to be connected with a multitude of spaces, both physically, as versatile individuals, and through various kinds of networks and contacts. And as people travel, migrate, expand their settlements and exchange goods, a concept like agricultural intensification is a slippery soap to handle. For example, when people migrate from a dry plain to a wetter mountain region, and in the process intensify production, it cannot be defined as intensification in a strict sense. Similarly, the observed abandonment of intensive farming practices, in association with the expansion of Iraqw agriculture to drier areas during the twentieth century, is better understood as local adaptations than a process of disintensification (see Sturdy 1936, Hartley 1938, Schultz 1971: 98-100). Agricultural intensification and disintensification are by definition processes over time in the same place, measured as either increased inputs or outputs by area unit and time (see Morrison 1994: 114-16, Netting 1993: 271, Turner et al. 1993: 10, Brookfield 1972: 31). Relative agricultural intensity between places, and even relative intensification as an effect of settlement expansion or migration, imply changes across space and are thus different from genuine intensification. Agricultural intensification must therefore be understood from a local context, and studied over time, *in situ*, in relation to an analysis of its wider, regional, socio-economic and political geography.

Apart from the need for spatial continuity, the crux of the population pressure theory is the assumption that people strive to maximize their labour efficiency (see Morrison 1994: 130-5, Bronson 1972). Since agricultural intensification, as a substitution of labour for land, requires increased inputs of labour per area unit and time, it is normally assumed that intensification reduces the marginal productivity of labour, thus resulting in diminishing returns on invested labour (see Boserup 1965). In other words agricultural intensification by means of labour power is generally regarded as a laborious process, which farmers are forced into by necessity. However, a distinction may be made between innovation and intensification (Brookfield 1984: 16): innovation takes place when farmers adopt new crops, tool or techniques, while intensification refers to the intensified application of inputs 'within a particular technological and socio-economic system'. According to this definition innovation is needed to increase labour productivity. But, for the purpose of this discussion one can, like Boserup (1965), consider such innovations as part of an intensification process (and even Brookfield acknowledges the frequent difficulty of drawing the distinction in practice). Hence, one can imagine situations where intensification may actually pay off as increased returns on marginal labour input, or at least without any marked decrease in marginal labour productivity, for example in the case of certain soil-conservation measures. In a case study from Indonesia, Padoch (1985) has shown that intensive permanent rice cultivation is a more efficient and more reliable subsistence technology than extensive shifting cultivation, despite a low population density. Depending on the particular interaction of social and natural factors in the area, intensive farming practices were adopted at an early stage in the colonization process, when land was available for expansion and extensive shifting cultivation. In fact, Padoch envisages the possibility of increasing population densities resulting in greater dependence on shifting agriculture, as more marginal areas need to be appropriated for agriculture (p. 287). Moreover, Bronson (1972) argues that labour-intensive farming is often associated with high labour productivity and that intensive farming practices have frequently been able to compete with more extensive forms of production, such as swiddening and long-fallow cultivation. It has also been shown that intensification may occur as an incremental process, where landesque capital is formed through the process of cultivation (Doolittle 1984), that is without any significant extra labour input. The process of intensification is hence not always associated with thresholds of labour demand.

Thus the adoption of intensive farming practices is not necessarily associated with a long-term evolutionary process of intensification, occurring in the wake of population pressure. The tendency, in much literature on African agricultural history, to treat intensive farming as a late state in an evolutionary sequence, without support from empirical data, must therefore be attributed to a lax use of theory. A common shortcoming is the failure to distinguish between high population density and high population pressure on natural resources, as if they have always occurred simultaneously. But, if population pressure is understood as a lack of resources, there is plenty of historical evidence which shows on the contrary, as does the Iraqw case, a correlation between concentrations of agrarian populations and sustainable labour-intensive farming without any notable pressure on resources. This is obviously more or less what Boserup (1965) argued, but when she concluded that population growth ultimately leads to technological innovation and agricultural intensification, she recognized scarcity or pressure on resources as a main driving force. The question is therefore to what extent scarcity or pressure is responsible for the type of intensification discussed by Boserup. In the Iraqw case the assumption that high population densities necessarily involve heavy pressure on resources has been more or less axiomatic in earlier literature. Thus, in competition with population pressure theory as an explanatory model of agricultural change, empirical evidence of Iraqw agricultural history has, till now, been accorded little importance.

Although it is clear that the Iraqw expansion was largely facilitated by colonial policies and action (Lawi 1999), and not just a simple response to population pressure within the homeland, there is still no satisfying historical explanation of why and how Iraqw intensive agriculture evolved. By a critical examination of Iraqw historiography, complemented with the use of landscape illustrations and photographs, this chapter both highlights the ambiguities of the historical source material, and attempts to produce historically grounded conclusions about precolonial and early colonial Iraqw settlement history.

The inspiration to tackle the historical problems associated with Iraqw intensive agriculture in the Mbulu Highlands has derived from investigations of landscape history and agriculture carried out by the author in a small section of the Iraqw homeland, intermittently between 1996 and 1998. The focus of this fieldwork has been a detailed participatory mapping of the history of land use and land tenure in the area surveyed. The mapping of physical landscape features has thus been juxtaposed with the stories told by local farmers (Börjeson 1998, 1999). This fieldwork methodology has been outlined in earlier publications as a way of 'listening to the land' by stressing the use of the landscape as well as oral history as source material. With the help of Geographical Information System (GIS) software, historical maps of the surveyed area can be produced, based on interview data and aerial photographs. A map of a selected area of the present agricultural landscape is shown in Figure 4.2 (see also Figure 4.3). Although the analysis in this chapter is based mainly on colonial and secondary source material, it also relies, but in an indirect way, on the insights gained from this fieldwork. The landscape is still in focus, but the level of geographical resolution is much more general in the analysis pursued here, compared with the fieldwork data. However, as far as possible the intention of 'listening to the land' has been upheld by matching historical evidence of physical landscape features with colonial and local observations of landscape changes.

Agriculture in Iraqw'ar Da/aw

> *Siník!* (Here is a riddle!)
> *Geerankáy!* (Open it up!)
> *Hiriki 'Umbúuru?* (The undulating of Mbulu?)
> *Doosla.* (Agriculture)
>
> *I /iif ku gagaara 'úr?* (It is very big and too heavy to carry?)
> *Kunti.* (The granary)
> (Two Iraqw riddles recorded by Paul Berger in 1935, and reproduced in Berger and Kiessling 1998: 294-5, 297-8)

The Iraqw homeland roughly conforms to Kainam and Murray wards in what is now designated the Endagikot Division of Mbulu District. In the Iraqw community the area is referred to as Iraqw'ar Da/aw, but the names

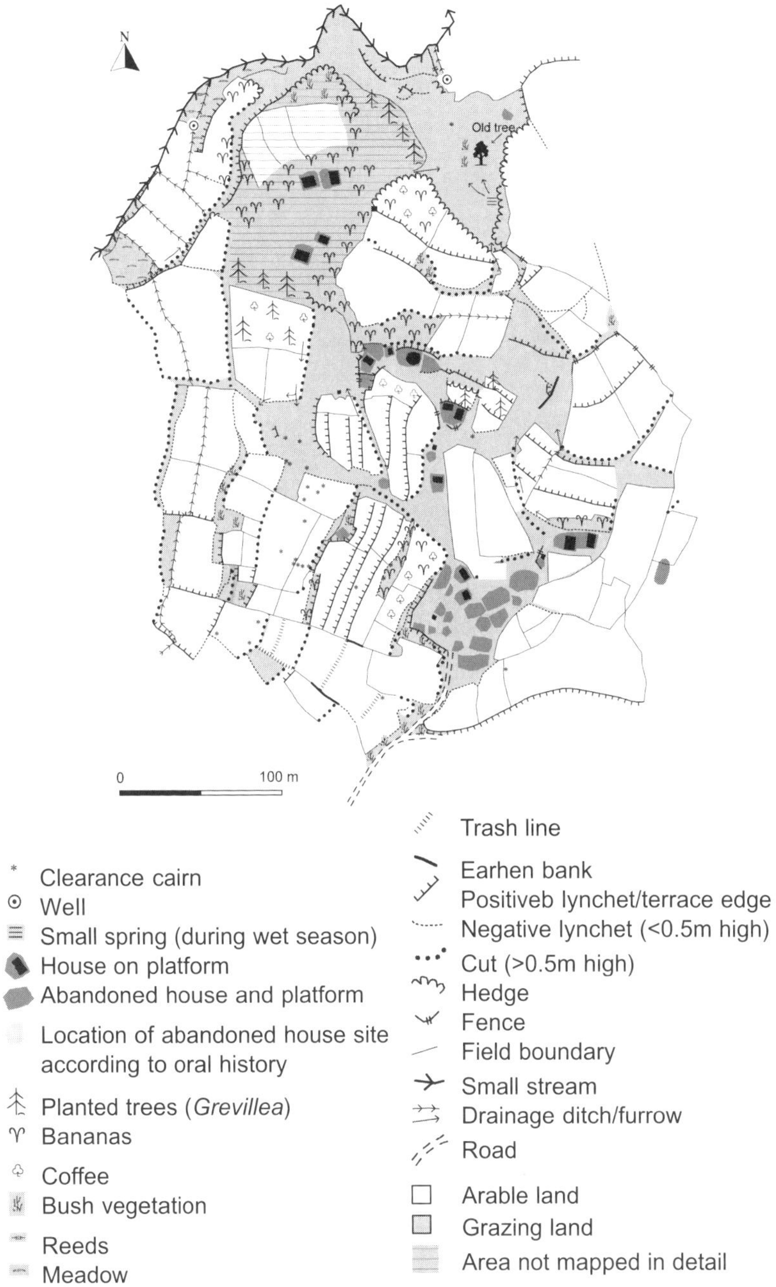

Figure 4.2 *An agricultural landscape in Hhay Geay, Iraqw'ar Da/aw (mapped by Lowe Börjeson and Deogratius Hillu, 1996, 1998)*

Mama Issara and Kainam are also commonly used. The name Iraqw'ar Da/aw, meaning the eastern Iraqw land or the eastern Iraqw people, is of particular interest since it indicates a spatial relation. Lawi suggests that the first Iraqw settlement in the area was simply named Iraqw following the common practice of using the same name for places and peoples (Lawi 1992: 47). Hence, the name Iraqw'ar Da/aw implies a migration or expansion process, but the actual direction of this movement of people (in or out from the Iraqw'ar Da/aw area) or when it took place is more difficult to ascertain.

During the twentieth century Iraqw'ar Da/aw became economically marginalized in Iraqw society as a result of the expansion of settlement. In 1988 the population of Iraqw'ar Da/aw was estimated at just over 19,000 people, compared with the total of about 500,000 Iraqw living in Mbulu and neighbouring districts. Population growth in Mbulu District is as high as 3.8 per cent, one of the highest rates in the country. But in Iraqw'ar Da/aw the rate of population increase during the twentieth century was lower because of the continuous out-migration (Snyder 1996: 318, 324).

The area is located in the highlands only a few kilometres south of Mbulu town at an elevation ranging from 1,500m to 2,300m, with the main cultivated land at around 1,800m. The area is bounded by a fringe of hills, which in the southwest, west and east are covered with mountainous forest in two forest reserves. In the southeast the area is delimited by the Manyara Rift escarpment. Between the cultivated area in the centre and the surrounding higher hills lies a zone of extensive grazing land, which is mainly covered with bracken. The landscape is completely dissected and dominated by ridges forming a dendritic pattern with elongated valleys between. From the forested hills encircling the area a number of permanent rivers meander through the larger valleys of the area. Small springs are found in nearly every valley, creating a network of smaller streams, many of which are permanent. The bedrock is deeply weathered soft gneiss, and the soils are of moderate to low fertility on the slopes, varying from low to high in the valley bottoms (National Soil Service 1994, III: 13-14, 63-4). Generally, however, soils in the upper parts of ridges are shallow and of low to occasionally very low fertility, while lower-lying slopes and reasonably well-drained valleys contain clay soils of moderate fertility (Meindertsma and Kessler 1997). The mean annual rainfall is high, up to 1,000mm or more, but varies greatly.[2] However, evaporation is relatively low owing to the cool and often foggy conditions of the area (NSS 1994, I: 23, Schultz 1971). The months between June and September are usually the driest months with little or no rain (Meindertsma and Kessler 1997: 12). These natural endowments, particularly the availability of water, make the area suitable for intensive agricultural production, despite the limits imposed by the dissected topography, leached soils and the variability in rainfall.

Farming in the area is a year-round undertaking. In every season there are crops to be planted or harvested and fields to be prepared, not to overlook all the necessary weeding, thinning and pruning, etc. On the rain-fed slopes crops are frequently rotated, and on many fields more than one crop is harvested each year. Wetter valleys may produce crops even during prolonged

Figure 4.3 *Agricultural landscape in Iraqw'ar Da/aw, showing the northwestern part of the area mapped in Figure 4.2, looking southwards (Photograph: Lowe Börjeson)*

droughts. Mechanization of production is difficult and also risky in this hilly and erosion-sensitive environment. Cultivation is consequently done by hand. Land is divided into four major categories: crest (*dindirmo*), eastern slope (*intsi*), western slope (*genei*) and valley (*khatsa*), which are all regarded, by Iraqw farmers, as having unique properties. Various crops are grown: maize, beans, sweet and Irish potatoes, sorghum, finger- and bulrush-millet, wheat, cassava, various vegetables and fruits, tobacco and coffee. Maize is the main staple, and cash crops like coffee and tobacco are grown in small quantities and are partly used for home consumption. Valley-bottom fields are mainly intercropped with maize and beans, while cropping diversity is much greater on the slopes. Fields are small, 0.1–0.3 ha, and an average farmholding is about 1.4ha, including a small home-grazing area (Snyder 1996: 321). Cattle, goats and sheep are kept, and pigs are raised for the urban markets in Arusha and Moshi. Only a few improved cattle are found in the area. The local zebu are usually grazed during the day and kept inside the houses during the night. Wet manure from the night is dried in the courtyard during daytime and returned to the house as bedding. This procedure is repeated a few times before the dung is put on a manure pile for spreading later on the fields.[3] Apart from cow-dung, various types of compost and green manure are used to restore the fertility of arable fields, as well as short- and long-term fallow. The input of organic fertilizer is not only a means of restoring nutrients, it is also an active interference with soil formation processes. Soil quality is not simply a constraint to agriculture in Iraqw'ar Da/aw, it is also a product of farmers' cultivation practices over time.

Figure 4.4 *Typical cut (or terrace edge), approximately 1m high, Hhay Geay, Iraqw'ar Da/aw (Photograph: Lowe Börjeson)*

While hoeing the slopes in the area, farmers move soil downhill. After many years of cultivation, slopes therefore become gently terraced, as cuts develop on field boundaries (Figure 4.4). The upper parts of the cuts, or the terrace edges, are frequently enhanced with grasses, trees, bananas or bushes, and storm drains are sometimes dug to lead off excess water. Crops are usually planted in ridges across the slope. Frequently, crop residues are buried in ridges or distributed between ridges as mulch. Lines of crop residues (trash lines) are also laid out to prevent erosion. Many valleys have been drained for cultivation and furrows are often needed to prevent waterlogging. When planting takes place under dry conditions, seeds are often watered by hand to facilitate germination. Occasionally, plants are also watered as a protection against crop damage caused by cutworms (*Agrotis ipsilon*).

Yields appear to be moderate or relatively low. Estimates of maize yields are 0.7–2.7 tonnes/ha (Snyder 1996, Meindertsma and Kessler 1997: 42, NSS 1994, I: 31, Liwenga and Lyimo 1996). Local farmers claim that yields have declined in the last few decades, especially on the slopes. Though there seems to be no simple explanation for this assumed decline in productivity, recurring droughts, insufficient organic fertilizers and less labour availability are some possible reasons voiced by farmers (Snyder 1996: 323-4, Liwenga and Lyimo 1996). According to the estimated yields, agricultural productivity in Iraqw'ar Da/aw is low in terms of output per hectare when compared with other areas in the district where farming is more mechanized and capital-intensive. But considering the difference in soils, climate (lower temperature and more cloud) and investments in soil quality and also the various crops grown and the types of intercropping, multicropping and crop rotations practised in Iraqw'ar Da/aw by comparison with other parts of Mbulu District, these measures of maize yields are of little specific value. Hence, as a measure of intensity, any attempt to calculate the output (yield) per area unit and time is hopelessly problematic in this diverse and complex farming system. However, the careful soil- and water-conservation practices described above, as well as the small farmholdings, clearly indicate a high level of labour intensity, which in turn increases area productivity.

Houses are generally on the upper part of the ridges and each household has its own small and well-maintained, home-grazing area around the homestead. Groups of households, moreover, share additional grazing land nearby, both in valleys and on ridges, which are carefully managed in order to ensure grazing round the year. Large, communal and rougher grazing land is also available outside the main settled zone, including forest glades. Wet meadows are also common. These are valley-bottom fields with evergreen vegetation, where grazing is usually not allowed and the grass is cut and fed to the animals by hand. Their main function is as fodder reserves for cattle with special needs, such as sick animals, calves and brood cows.

Roads to and from the area are very poor and there is no reliable car or lorry transport. Goods are transported on foot, by bicycle or on donkeys. This makes the area rather isolated from major markets, even though Babati and the busy road between the major towns of Arusha and Singida are not, in fact, very far away.

As in most parts of the world, agriculture in Iraqw'ar Da/aw underwent conspicuous changes during the twentieth century. The introduction of pigs and a number of crops such as potatoes, wheat, coffee, and various fruit and timber trees, and equally the increasing reliance on maize as a staple, together with a general shift from wooden implements to iron hoes and an expansion of valley-bottom cultivation, have all affected the appearance of the Iraqw'ar Da/aw landscape and created new opportunities for production. However, not all current farming practices in the area were new to the twentieth century. For example, labour-intensive practices, such as manuring, mulching and construction of ridges and storm drains, appear to have a longer history in the area, as discussed below.

Aspects of Iraqw settlement history

> 'The existence now … of Southern Cushitic-speaking cultivators with cattle (Iraqw, etc.) should not be seen as a simple relic of the situation before the expansion of both Nilotes and Bantu … The emergence of the Iraqw as such and their successful agricultural adaptation to the hills of Mbulu district are very recent developments.' (Sutton 1993: 51-2).

The Iraqw are the descendants of a multitude of ethnic groups now living in the region around Mbulu District (Thornton 1980, chap. 8). However, as part of the Southern Cushitic language group, the Iraqw are associated with a long history of agro-pastoralism in the region of the eastern Rift Valley, stretching several thousand years back (Sutton 1993: 51-2, Ehret 1998: 10, 31, Ambrose 1982: 113). Moreover, the ethnic landscape is continually changing, as new economic, political and religious opportunities develop and others are marginalized. Among the Iraqw this is particularly evident in the wake of their territorial and cultural expansion during the twentieth century (Rekdal 1999). Although one can discern an expansion of Iraqw culture outwards from its precolonial core area during the twentieth century, at the same time one comes to appreciate how that culture has also transformed and diversified itself in the process, encapsulating a range of what may be variously labelled as modern and traditional values. The sense of being Iraqw is a different experience for a Datoga trader, who may aspire to that identity to promote his or her business in the towns of Mbulu or Karatu, from that of an Iraqw elder, who needs to make long journeys from his home area to Iraqw'ar Da/aw to promote clan and ritual relations (see Rekdal 1999: 39-44, Loiske n.d.).

From the evidence of genealogies as remembered (Fosbrooke 1955: 17, Thornton 1980: 193-4, Lawi 1999: 19), the date of the settlement of Iraqw'ar Da/aw has been estimated as no later than the end of the eighteenth century. However, even this rough date must not be taken too seriously, especially not in the sense of a primary colonization of this highland area. It is possible that the area experienced much more ancient human settlement and farming.

Moreover, Iraqw genealogies collected by Loiske (in 1995–6) reveal considerable uncertainty among informants about the sequence and number of forefathers that constituted their clan. One of the problems was that sons were often named after their grandfather, so that names were repeated every second generation. Frequently informants felt a need to consult others in the clan in order to offer a correct genealogy. This procedure commonly resulted in a consensus compromise that was acceptable to the clan. It was consequently difficult to distinguish between a clan and its component lineages, and also to discern the number of generations represented by these genealogies (Loiske, pers. comm.).

The Iraqw myth of origin tells of a move caused by warfare or drought (or both) from a place in the south to the Mbulu Highlands.[4] But considering the multi-ethnic origins of the Iraqw, it is probably unrealistic to attempt to locate any previous Iraqw homeland. A more complex picture of precolonial Iraqw migrations is given in an Iraqw historical narrative collected by Thornton (1980: 160-4). Equally doubtful would be the existence of any discrete Iraqw ethnicity before their settlement in the Mbulu Highlands. Most likely, however, the colonization of the Iraqw'ar Da/aw area was a gradual process, involving clearing highland forest, which would have started long before the Iraqw dominance (see Lawi 1999: 21-4). In other parts of East Africa, mountain forests have been used and successively colonized during a much longer time than that covered by Iraqw genealogies, and such forests have decreased owing to expanding cultivation. The Mbugu, a Southern Cushitic pastoral people long resident in the Usambara Mountains, is one parallel example (Conte 1996).

The increasing number of Iraqw clans clearly indicates the extent of cultural assimilation in the history of this people. Of the 150 or possibly 200 Iraqw clans,[5] only three are regarded as being of Iraqw origin (Snyder 1993: 97, Lawi 1999: 19).[6] Overall, the implication of the oral evidence is that an Iraqw culture and identity primarily developed during the nineteenth and possibly eighteenth centuries. In this respect people of Nilotic and Bantu derivation comprise an integral part of this culture and identity; they cannot be regarded as mere immigrants.

Regional, historical and environmental conditions for agricultural production in Iraqw'ar Da/aw

Iraqw'ar Da/aw lies between the two main caravan routes which linked Lake Victoria and the coast in the nineteenth century; these passed approximately 100km to the south and north (Koponen 1988a). Although one should probably imagine a much more complex and wide-reaching transport and trade network, Iraqw'ar Da/aw appears to have been, if not isolated, certainly on the margins of late precolonial economic development in Tanzania. Beside connecting the East African interior more directly with a global economy, this intensified long-distance trade carried new infectious diseases, slave-raiding and warfare in its wake (Kjekshus 1996, Koponen 1988a). It may be assumed, therefore, that most of this disruption and ruin did not affect Iraqw'ar Da/aw in as direct a way as was experienced in other areas. There is also a common belief among the Iraqw that the slave trade was successfully kept at a distance by means of ritual power (Snyder 1993: 17). However, the resultant turmoil in badly struck areas stimulated people to move towards more remote places (Koponen 1988b). To what extent such demographic turbulence in the late nineteenth century hit Iraqw'ar Da/aw is unclear. Bagshawe (c. 1920: 37, 1926) notes that the Iraqw suffered from famine and smallpox before a period of settlement expansion, which suggests disease and hardship were not absent in Iraqw'ar Da/aw in the late nineteenth century. Overall, however, it seems that the area was among those which were only marginally affected by the worst calamities of that time, and that this likely had a positive influence on settlement expansion in Iraqw'ar Da/aw. The in-migration to the area and the development of Iraqw society, as indicated by the growing number of clans, as observed, may thus have been related to the political, economic and demographic turbulence of the late nineteenth century.

Coercion or cohesion?

Previous accounts of Iraqw history strongly agree that the Iraqw were confined to a small and isolated area during the nineteenth century and that this led to mounting pressure on resources as the population grew, which in turn created a need to intensify agricultural production. There are basically two main explanations put forward for this isolation: the physical geography, with the surrounding mountains, forests and escarpment, and the supposed aggressions by Maasai and Datoga pastoralists who occupied and controlled the plains outside Iraqw'ar Da/aw. These two arguments contradict each other to some extent. First, the Iraqw culture, as it developed during the nineteenth century, was mainly consolidated through the amalgamation of people from surrounding ethnic groups, including Datoga and Maasai (Lawi 1999: 27). Secondly, the physical geography they encountered not only protected them from assault by the surrounding pastoralists, but also provided favourable opportunities for both agriculture and trade.

There are frequent references to Maasai and Datoga hostilities in Iraqw oral tradition (Berger and Kiessling 1998: 146-87, Lawi 1999: 34, 65-6). However, these hostilities appear to have varied in intensity between seasons. It is told how the Datoga negotiated for peace with the Iraqw every dry season in order to obtain food, for instance by trading cattle skins for millet (Berger and Kiessling 1998: 173-6). Some attacks ended in friendship and the creation of new Iraqw clans and, indeed, many clan histories describe how Maasai happened by chance to stay in Iraqw'ar Da/aw (Loiske, pers. comm.). A typical clan history may tell of a Maasai boy who came to Iraqw'ar Da/aw to steal cattle, but who, while spying on one of the neighbourhoods, saw an Iraqw girl with whom he immediately fell in love and therefore decided to stay in the area. It would be naive to interpret such stories as evidence that interaction and exchange between the two people were consistently peaceful, but their mere existence suggests that Maasai and Iraqw interaction was more complex than that of attacker and attacked. For example, how could an Iraqw family accept a fierce enemy as a son-in-law in the family, even if it is just a story?

The view that Iraqw society had been isolated and confined by coercive external force, what is here called the siege hypothesis, has drawn support from an argument formulated by Fosbrooke in the 1950s, to which subsequent accounts of Iraqw settlement history have frequently referred. Fosbrooke contended that the Iraqw changed their style of housing, from round houses with high conical thatched roofs to low partially subterranean and earthen-roofed *tembe* houses, as a defensive measure against Maasai raids during the early nineteenth century. These *tembe* houses, by this view, were thought to have been in use during a limited period only, after which the round-house style was adopted again (Fosbrooke 1954). However, as indicated by Baumann and Jaeger at the end of the nineteenth century and beginning of the twentieth century respectively, *tembe* houses were normally constructed in the areas around the periphery of Iraqw'ar Da/aw, and round, high-roofed houses were common inside the central parts of the homeland (Baumann 1894: 118-21, Jaeger 1913: 53-4; see also Thornton 1980: 123, Berger and Kiessling 1998: 186-7; see Figures 4.5-4.9). In fact, Jaeger refers to a particular 'area of round houses' (*Rundhüttengebiet*) inside Iraqw territory (see below). But even if this does not strictly disagree with Fosbrooke's argument that *tembe* houses were constructed in Iraqw'ar Da/aw earlier in the nineteenth century, there is no substantial evidence suggesting that *tembe* houses were adopted on such a grand scale and for defensive purposes, as he suggests. The difference in house styles inside and outside Iraqw'ar Da/aw, as noted by Jaeger, may thus have been a much more persistent feature in the precolonial settlement history of the area, a difference which may be explained, in part at least, by variations in local climate and the availability of suitable building materials. Fosbrooke's argument seems to have been over-influenced by his assumption that the Iraqw were under attack by the Maasai, which in turn makes it inseparable from the siege hypothesis itself.

Hence, the formation of an Iraqw society during the nineteenth or possibly eighteenth century appears to be more related to processes of social and political cohesion than to external coercive pressure. A more valid picture might see Iraqw'ar Da/aw as a cultural melting-pot, where immigrants quickly adopted an Iraqw way of life and adherence to Iraqw authorities. Welcoming and inviting immigrants is a common theme in Iraqw oral history. Indeed, immigrants could quickly achieve important positions in Iraqw society, as ritual leaders for instance (Lawi 1999: 28-9). In precolonial times Iraqw society was by and large undifferentiated, but ritual leaders and respected elders (mainly men) had dominant social and political positions as well as economic power (ibid.: 34-5). Acceptance of these authorities probably contributed to cultural cohesion (Thornton 1980: 189-211), but it was not the only factor. The presence in the surrounding area of mutual enemies, both Maasai and Datoga (Lawi 1999: 29), is evident, but also environmental factors may have contributed to the shaping of an Iraqw identity and culture, and stimulated settlement concentration (Lawi 1999: 28-9).

Relief and opportunity

As a well-watered highland area Iraqw'ar Da/aw has been able to help surrounding communities with relief during periods of sustained crop failures or lack of fresh grazing. Broadly, the situation has been similar to that of the Usambaras, Pare and Kilimanjaro Mountains in north-east Tanzania, which have for millennia attracted human settlement that has expanded in times of drought (Conte 1994: 29-56, Kimambo 1996, Hakansson 1995). The natural endowments of Iraqw'ar Da/aw – high rainfall, moist and cool climate, steep slopes, wet valleys and forests nearby – would, moreover, have played a crucial role in influencing the organization and structure of agricultural and animal production. The high groundwater table in the valleys provides a kind of natural irrigation if excess water can be drained. This has assured fresh grass round the year close to the homestead, and thus opportunities for intensive livestock production.[7] Cattle were also kept at home during night-time for protection (mainly from wild animals) and convenience (Lawi 1999: 61-2, 104). Keeping the cattle indoors at night also increased warmth and comfort during damp and cold weather. Manure was hence readily available in the household compound and could easily be spread on the nearby fields to enhance fertility and boost production. The altitude also provides relief from the deadly parasites carried by tsetse fly and malaria mosquitoes.

Despite the distance to major market places and caravan routes, the Iraqw were engaged in trade and exchange with neighbouring peoples during the nineteenth and early twentieth centuries. This trade was most intense with the Mbugwe and Gorowa, living below the Rift wall. But they also exchanged goods with the neighbouring Datoga, Nyaturu, Nyiramba, Isanzu and lately also with the Gogo (Lawi 1999: 36-7, 44, 65). Regardless of whether this is labelled as trade, exchange or some other form of socio-economic relations, such networks spanned the region, connecting dry and wet, high and low areas into a regional economy that is still of great importance to the Iraqw (Thornton 1980, Lawi 1999, Loiske, Chapter 5). Hence the economic and agricultural opportunities provided by the location and environment of Iraqw'ar Da/aw render the siege argument superfluous to the analysis of Iraqw settlement history.

Agricultural intensity in the nineteenth and early twentieth centuries: the evidence of observers and photographs

No archaeological or palaeo-botanical and environmental evidence specific to the early settlement history in Iraqw'ar Da/aw is available, and, apart from local oral accounts, eye-witness descriptions from the late nineteenth century are few. The available historical material is thus quite meagre, but at least some clues about the character of the landscape and its utilization can be found in early travellers' accounts and later colonial reports. Three German expeditions passed through the area around the turn of the nineteenth century, those led by Baumann (in 1893), Werther (in 1896–7) and Jaeger (in 1906–7). These descriptions of the landscape and agriculture are accompanied by photographs or prints (in Baumann and Jaeger).[8] A later article on the agriculture of Iraqw'ar Da/aw (Hartley 1938) includes landscape photographs.[9] By analysing the landscape shown in these pictures in relation to the written descriptions, the extent and character of Iraqw settlement and agriculture from the late nineteenth century can be outlined.

Both Baumann and Jaeger were impressed by Iraqw agriculture. Jaeger describes the Iraqw country as a friendly 'cultural oasis' (*Kulturoase*) with an idyllic landscape (Jaeger 1913: 46). And Baumann admired the wide stretches of open land with its nicely kept arable fields, shining like red rectangles in the midst of the grass-covered slopes (p. 118).

After climbing the Magara escarpment from the Mbugwe plains south of Lake Manyara, Baumann's expedition entered Iraqw territory south of the Marang forest, passing through present-day Gehandu ward to the north of Iraqw'ar Da/aw. His journey continued around the Guwangw mountain massif to the area where Mbulu town now stands, and then south through Iraqw'ar Da/aw to 'Meri', which lies about 10km to the south, overlooking the escarpment. Judging from the extensive patchwork of fields stretching up

Figure 4.5. *'Iraqw landscape' (Baumann 1894: 119), which shows the mountain of Guwangw, which lies just beyond the northern border of Iraqw'ar Da/aw and a few kilometres east of Mbulu town. It is an eye-catching mountain, looking essentially as depicted when viewed from the west. There is no direct reference in Baumann's text to this mountain, but since he speaks of a 'proud mountain top' while passing through the very area, the identification is reasonably certain* (*Baumann 1894: 118-19*).

Guwangw mountain and the number of houses, as illustrated by Baumann (Figure 4.5), it seems that this side of the mountain was experiencing quite intense land pressure. The concentration of fields and their distribution in relation to the houses and topography, and a few larger blocks of possibly subdivided fields, indicate that this agricultural landscape was already of some age when Baumann passed by in the early 1890s. Although the area depicted lies outside the borders of Iraqw'ar Da/aw, it was perceived as a typical Iraqw landscape by Baumann, kept clear by agro-pastoral activities and dotted with low rectangular, flat-roofed *tembe* houses. It is worth noting that although Baumann passed through Iraqw'ar Da/aw, he did not clearly differentiate between this area and the rest of the Iraqw country, as Jaeger did a few years later (see below).[10] Baumann apparently considered that Iraqw'ar Da/aw was not a core area of Iraqw settlement, and his description of the Iraqw landscape is mainly concerned with areas outside Iraqw'ar Da/aw.

Figures 4.6a, 4.6b *'Northeastern Iraqw, at Dafi's. The mountains in the background form the crest of the great escarpment, hence the markedly dissected landscape. Fields and scattered* tembe *huts cover the slopes' (Jaeger 1913: Plate V). The area shown in the photograph, taken in 1906/7 on Jaeger's expedition, is clearly outside Iraqw'ar Da/aw. (On Jaeger's map Dafi's is about 8km northeast of Mbulu town, on the northern side of Guwangw mountain in present-day Gehandu Ward.)*

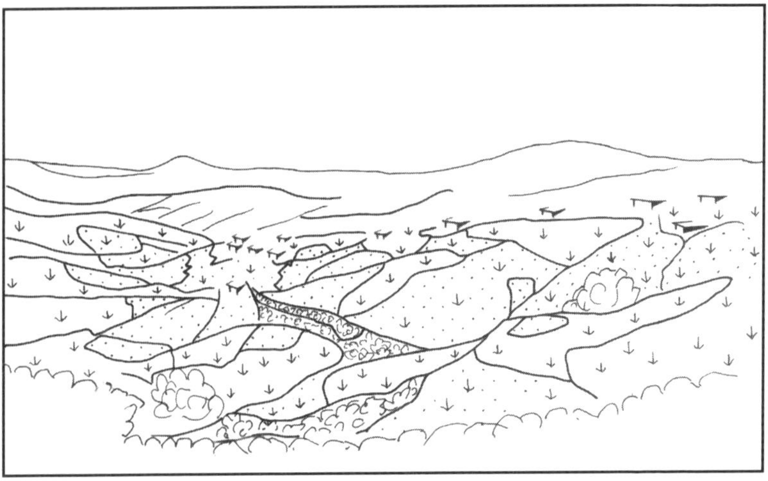

Northeast of Guwangw mountain, that is, further north of Iraqw'ar Da/aw, the land appears to have been equally densely settled and as intensively cultivated as on the slopes of Guwangw at the beginning of the twentieth century (see Figure 4.6). According to Jaeger this place lay close to the escarpment and the plains below (Jaeger 1913, Plate 5, Map 6, see Figure 4.6a,b). It is questionable whether this was an early Iraqw expansion, caused by population pressure in Iraqw'ar Da/aw, or was a settlement founded in a favourable location for both agro-pastoral production and the potential for trade and exchange. In any case this area must have been settled well before the date of Jaeger's journey (1906/07).

A small section of the landscape of Iraqw'ar Da/aw at the turn of the century is pictured in Figure 4.7. What Jaeger calls South Iraqw probably corresponds to Iraqw'ar Da/aw. The main purpose of the picture appears to have been to illustrate the distinct house type (round houses) in Iraqw'ar Da/aw, by contrast with the *tembe* houses in other parts of Iraqw territory. We may also assume that the landscape shows scenery typical of the central Iraqw'ar Da/aw settlement in the early twentieth century. Jaeger describes this southern part of Iraqw land as an enclosed area where there were no *tembe* houses, but exclusively round houses with conical roofs. Only at the outer rim of this area were people living in *tembe* houses (p. 54). He also notes that the northwestern part of this 'area of round houses' (*Rundhüttengebiet*) stretches out along the main road through the mountain pass between Nou forest and the Guwangw mountain, and into the area of *tembe* houses (that is, into what is now the southern outskirts of Mbulu town). Thus the borders of this 'area of round houses' match the present northwestern and eastern

Legend to interpretations of photographs, Figures 4.6b-9b

Land under cultivation (or short fallow)

Fallow field

Uncertain: cultivated, fallow or grazing

Grazing land

Bush vegetation

Unidentified field structure

Cut/terrace edge

Round house, with high conical thatched roof

Tembe house, with flat earthen roof

Figure 4.7a, 4.7b *'South Iraqw: round houses at Issara's, with ferns' (Jaeger 1913: Plate V). Picture taken at a place in east-central Iraqw'ar Da/aw, around the present border of Kainam and Murray wards (if correctly plotted on the map by Jaeger: 'Issara's' on his map). In the foreground there is a stand of ferns, then a valley field followed by a gentle slope, with fields and patches of grass leading up to the home-grazing area around a cluster of four houses. In the distance the landscape is open and no trees are in sight.*

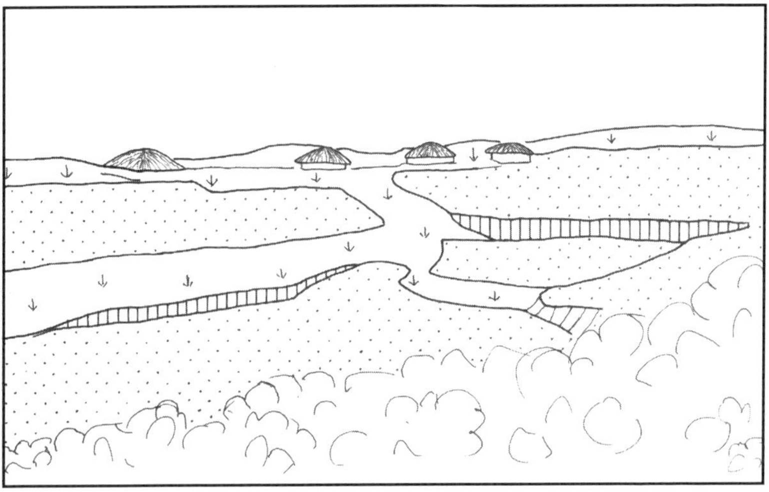

borders of Iraqw'ar Da/aw quite well. Hence Jaeger identifies Iraqw'ar Da/aw, contrary to Baumann, as a place with a particular physical geography and characteristic settlement.

Equally important in Jaeger's description of Iraqw territory is its extent (Jaeger 1913: 53). He portrays it as a well-populated basin surrounded by mountains and wilderness. In the centre of this confined area stands the conspicuous Guwangw mountain, around which the cultural landscape stretches out in all directions. Nowadays this area would include roughly the lower-lying areas around Mbulu town, to the west, north and northeast of Guwangw mountain massif, up to the Marang forest, and the higher Iraqw'ar Da/aw area on the other side of Guwangw mountain. In addition there was an area settled by the Iraqw around Lake Tlawi (Bassoda Merka on Jaeger's map, 1913: Map 6). Hence, Jaeger's reference to the Iraqw land as a densely populated, agro-pastoral landscape, appearing as an oasis surrounded by wilderness, applies not only to Iraqw'ar Da/aw. This means that his observations of good agricultural practices among the Iraqw, such as manuring, irrigation and crop rotation (Jaeger 1913: 54), would also pertain to areas outside Iraqw'ar Da/aw, implying that the intensity of cultivation and population density was high there also at the very beginning of the twentieth century. For this reason, and from the above observations, it may be concluded that a large part of the Mbulu Highlands, not merely Iraqw'ar Da/aw alone, must have been more or less permanently settled by agro-pastoralists (not necessarily to be identified as Iraqw) from at least the mid-nineteenth century, if not earlier.

Farming took up a large proportion of the land in the Iraqw enclave, according to Jaeger's observations. In good places up to half consisted of arable fields, leaving little space for grazing.[11] However, south of Mbulu, he claims to have seen several square kilometres covered almost entirely by arable fields, apart from the *tembe* houses and roads. Significantly, Jaeger here appears to be speaking of the area just south of the German headquarters at present Mbulu town, north of Iraqw'ar Da/aw. In more peripheral areas, closer to the surrounding mountains, settlements are said to have been less dense with cultivated areas covering only about a tenth of the land (Jaeger 1913: 54). Although the small portion of Iraqw'ar Da/aw shown in Figure 4.7 is largely covered with arable fields, this is probably not representative of the Iraqw'ar Da/aw landscape in general, as arable fields would have been located near the houses. Beyond the frames of this picture there would certainly have been large tracts of uncultivated grazing land. However, it is obvious that the landscape is being sculpted to some extent by the farming practices, notably the levelling of fields and cuts developed on field boundaries in the slopes. The lower field, for example, appears to be shaped by this process, and thus must have been cultivated for some time. Thus, although there is evidence of labour-intensive farming practices in Iraqw'ar Da/aw, according to Jaeger some of the most densely populated and intensively cultivated area in 1907 was located north of Iraqw'ar Da/aw, in the lower Mbulu hinterland.

Furthermore, the print and photographs of Baumann, Jaeger and, later,

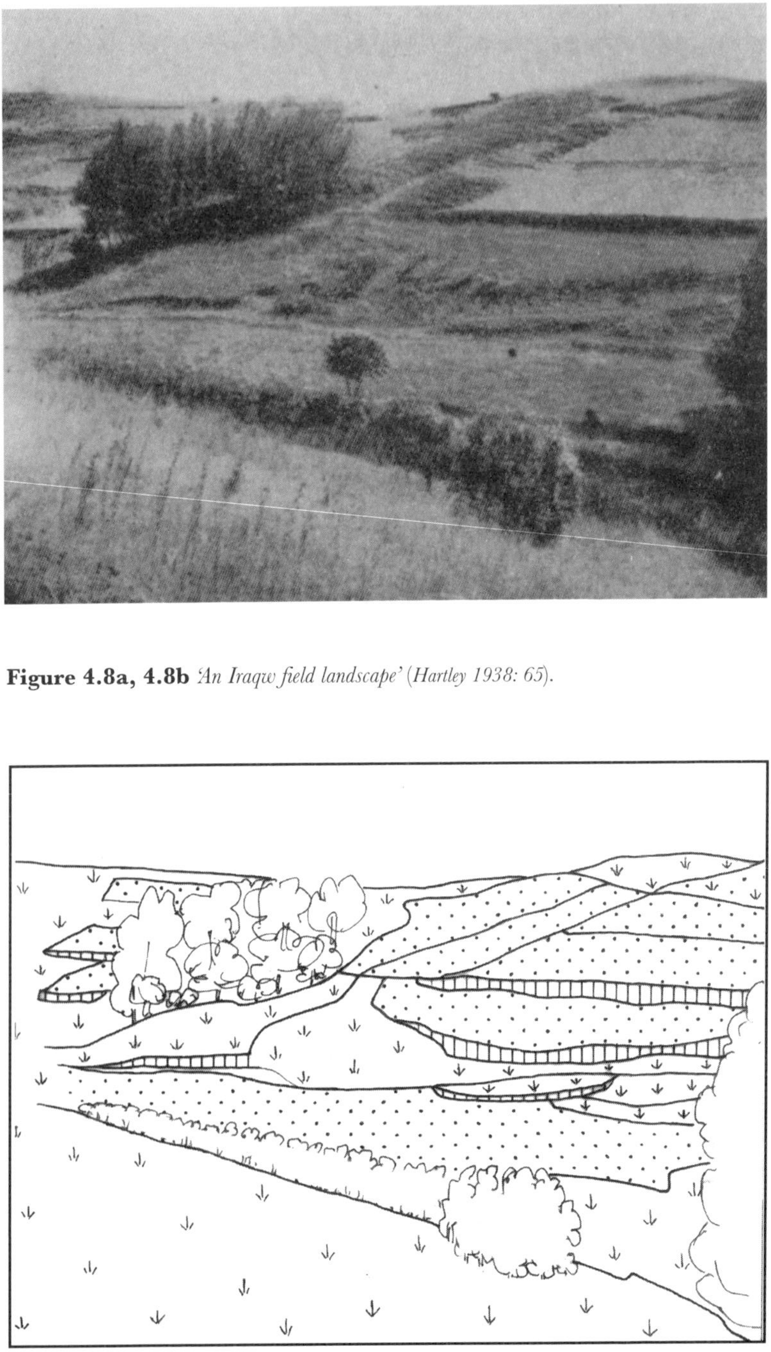

Figure 4.8a, 4.8b *'An Iraqw field landscape'* (*Hartley 1938: 65*).

Figure 4.9a, 4.9b *'Fields in the Kainam area, with crops planted in contour lines of field terraces'* (*Hartley 1938: 64*).

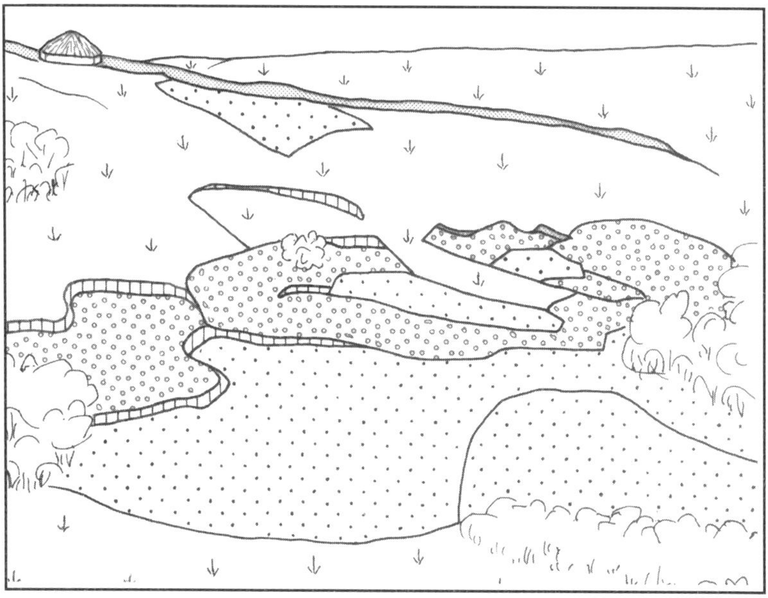

Hartley (see figures 4.5 – 4.9) show the existence of plenty of houses outside Iraqw'ar Da/aw, and the landscape inside Iraqw'ar Da/aw itself appears to be more sparsely populated (an observation consistent with the other pictures in Hartley 1938). Accordingly, cropping intensity, that is the amount of cultivated land in relation to fallow and grazing land, was similarly high in the areas outside Iraqw'ar Da/aw. In fact, the cropping intensity does not seem to be higher inside Iraqw'ar Da/aw. This impression may of course result from the selection of views and the photographers' purposes, notwithstanding that Hartley in the 1930s was specifically illustrating the labour-intensive farming practices in Iraqw'ar Da/aw. Figure 4.8a shows large parts of the exposed hillside are covered with arable fields, of which some were being cultivated and demarcated by several cuts, visible as darker lines. Just above the valley the vegetation seems rougher, with a patch of long-standing grass, which might imply low grazing pressure or a system of rotational grazing. Below are a couple of cuts or negative lynchets, demonstrating that the valley had been cultivated previously. The hill from where the picture was taken seems to be mainly covered with grass and bush. The stand of trees is most likely of recently planted exotic species. Note the absence of houses. Figure 4.9a shows fields with young crops appearing like patches in a landscape predominantly used for grazing. The roof of one round house is visible, but otherwise this landscape consists of open grassland with a few scattered bushes. A few fallow fields can be seen.

Beside the general contrasts already noted between the Iraqw'ar Da/aw landscape and that of the areas outside, in both physical geography and house style, further differences are illustrated in the pictures, notably in land-use patterns. In both areas houses are surrounded by grazing land as well as arable fields, but in Iraqw'ar Da/aw houses appear to be clustered in smaller groups with plenty of space between the clusters. The *tembe* houses outside Iraqw'ar Da/aw are more evenly spread over the landscape. So there seem to have been particular reasons for settlement concentration in Iraqw'ar Da/aw, which are not directly related to environmental factors or to a typical Iraqw culture. Instead, settlement patterns reflect local socio-economic adaptations, and thus in effect the practices and spatial organization of agriculture.

In the pictures of Iraqw'ar Da/aw physical field structures are visible, but it is difficult to determine if similar structures are present in the other pictures. The evidence of soil erosion (Figure 4.6) outside Iraqw'ar Da/aw is equally difficult to interpret, but it appears to be of a quite limited extent, related to some local factor, possibly livestock trampling, and is thus of little help in assessing the character of farming in the area. It is mainly from the impression of settlement density that we may comment upon the intensity of production, based on the print and photographs from areas outside Iraqw'ar Da/aw. Nevertheless, these pictures and the written accounts indicate that labour intensity (investment in soil conservation) was probably higher in Iraqw'ar Da/aw agriculture, but that cropping intensity may not have differed significantly between areas inside and outside Iraqw'ar Da/aw. One must, however, be careful of interpreting the differences in land-use and farming practices between Iraqw'ar Da/aw and its surroundings as a difference in settlement age, a result of a settlement expansion. In fact, there is no evidence in these pictures that settlement in Iraqw'ar Da/aw was considerably older than in those areas depicted outside the homeland. The differences noted can be explained by local adaptations to varying economic, cultural and environmental factors. Terms such as core and periphery or, indeed, homeland and frontier or expansion areas must be used carefully in relation to Iraqw settlement history.

The evidence of these historical eye-witness accounts and photographs agrees with the oral testimony collected by Lawi (1999: 93-7), indicating that the cultivated areas in Iraqw'ar Da/aw were on the slopes and towards the lower parts of the landscape, including valley bottoms, while ridge tops were reserved for grazing. Obviously, many valleys would have been reserved for grazing and thatching reeds, especially larger and wetter valleys with permanent streams or rivers. But some were most likely cultivated even before iron hoes became widespread in the twentieth century. Indeed, Lawi mentions that, according to local narratives and written sources, swamp farming may have been in place even before 1850 (1999: 106, n. 307). In the picture published in 1938 the Iraqw'ar Da/aw landscape is still essentially open, but bushes and stands of exotic trees are present. Hence the impression one gets from the pictures of Iraqw'ar Da/aw is of a fairly sparsely populated landscape, parts of which were intensively cultivated. Pockets of intensive land use (both cultivated fields and grazing areas) lay close to the homesteads. More extensively utilized areas, mainly for grazing, were interspersed between clusters of homesteads, and even more extensive communal grazing areas dominated on the peripheries of the settled area. In the late precolonial and early colonial period there would have been only a few densely populated areas in Iraqw'ar Da/aw (Lawi 1999: 98), concurring with Fosbrooke's observation in the 1950s: 'The area of Iraqw proper [Iraqw'ar Da/aw] could obviously be more heavily populated; the grazing in particular could be greatly increased in value by eradication of wire grass, a fact appreciated by the diviners who, with access to free labour, have cleared their holdings of this pest.' (Fosbrooke 1955: 53, cited in Lawi 1999: 99.)

Furthermore, as Fosbrooke's comment indicates, labour was not only mobilized from within households, but was also a matter of power relations, so that the capacity to intensify agriculture varied between households.

Estimates of population density

The siege hypothesis mentioned above is founded on the assumption that population pressure increases in the area under siege. In order to evaluate the validity of the siege hypothesis in the Iraqw case, it is necessary to discuss the historical estimates of population, which have provided empirical support for the siege argument. However, even if the data discussed here do not suffice for any realistic population estimates in an absolute sense, they may be used as a guideline for critical discussion of population trends. Further, by considering the spatial variation of settlement over time in relation to population data, a more contextual analysis of these figures is attempted.

Excluding surrounding mountains, forest and escarpment, Iraqw'ar Da/aw covers about 180 square km,[12] while the two existing wards of Murray and Kainam cover a slightly larger area. In the 1950s Fosbrooke reached a similar estimate, 170-80 square km, for what he regarded as the true cradle area of the Iraqw (1955: 28, 51). If one follows the siege hypothesis, this area would have been settled and cultivated to its full capacity towards the end of the nineteenth century, whereupon the Iraqw would have started to overflow into the surrounding areas (see Fosbrooke 1955, quoted on pp. 97-8 below).

Another source which indicates the extent of Iraqw settlement and farming activities around the year 1900 is a vegetation map published in Jaeger (1913). Figure 4.10 compares this map with two other vegetation maps of the same area, showing the situation in the 1960s and early 1990s. From this series of maps it is evident that the central area of open grassland in Iraqw'ar Da/aw expanded during the twentieth century. The driving force behind this change is most likely Iraqw settlement and agricultural expansion. In fact, it is likely that the open grass vegetation, as classified on each respective map, is more or less synonymous with arable and grazing land. This vegetation change may thus be used as a rough measure of the extent of Iraqw agriculture and grazing activities. Hence, there was probably plenty of room for expansion within the boundaries of Iraqw'ar Da/aw during the late nineteenth and early twentieth centuries.

On the vegetation map in Jaeger (1913) (see Figure 4.10), the open grass areas (interpreted here as cultivated and intensively grazed land) of Iraqw'ar Da/aw cover approximately 75 square km.[13] On Schultz's map (1971) the central open grassland covers about 120 square km, and on the most recent vegetation map of the area, central Iraqw'ar Da/aw is clearly visible as a mixture of crop- and grassland, which covers about 135 square km.

The growth of Iraqw population during the twentieth century resulted in settlement expansion both inside, as the above data show, and outside the boundaries of Iraqw'ar Da/aw. This means that if the Iraqw were really under siege during the nineteenth century they would have been enclosed inside a much smaller area than what is currently regarded as the homeland. Thus when estimating the size of an area such as the Iraqw homeland it has to be acknowledged that this refers only to a particular historical situation (or ethno-historical theory). The projection of a fixed homeland boundary back in time is one of the main weaknesses of the siege hypothesis.

Apart from the Iraqw cradle area, Fosbrooke also refers to the Iraqw homeland as the Kainam area (see quotation below). However, on his map of the region (1955, Fig. II), the area marked as Kainam covers nearly twice the Iraqw'ar Da/aw area, including most of Gehandu Ward north of the present Kainam Ward (see Figure 4.1). In fact, this area corresponds fairly well with the area of grassland mapped by Schultz and with the extent of

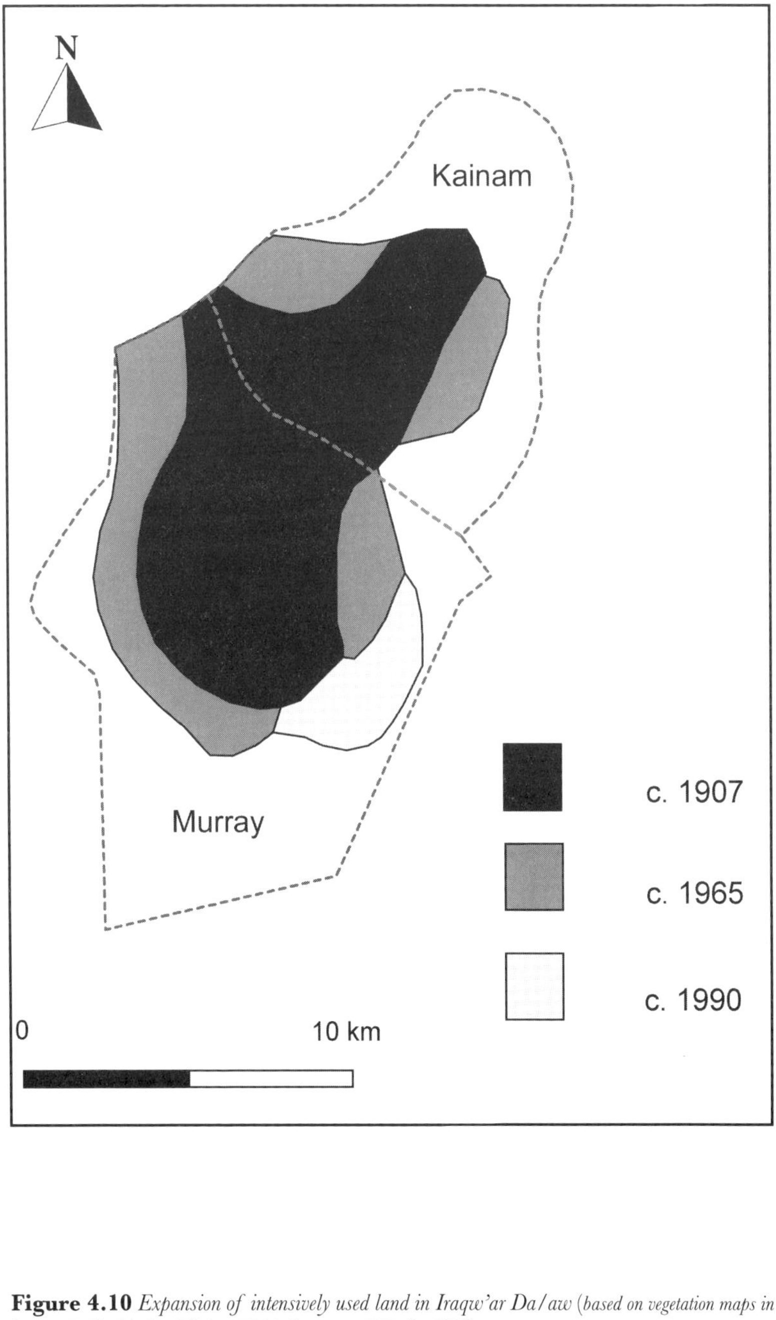

Figure 4.10 *Expansion of intensively used land in Iraqw'ar Da/aw (based on vegetation maps in Jaeger 1913, Schultz 1971 and Meindertsma and Kessler 1997)*

Iraqw settlement observed by Baumann and Jaeger around the turn of the century (excluding the settlements around present Mbulu town and Tlawi). This further suggests that the various references to the Iraqw homeland in colonial literature may in fact refer to different areas, and not necessarily to what is here delimited as Iraqw'ar Da/aw. One implication of this confusion is that in earlier literature Iraqw'ar Da/aw was often equated with the total area of Iraqw settlement during the late nineteenth century. This is another misapprehension which is part and parcel of the siege hypothesis.

Population figures and attempts to estimate density at various dates in the twentieth century are tabulated below. Densities are based on an area of 200 square km (unless otherwise indicated). The reason for this compromise figure is both the uncertainty of the actual areas pertaining to the population figures below and the difference between the area of Iraqw'ar Da/aw and the two present wards.

Population data for Iraqw'ar Da/aw

From the figures in Table 4.1, it seems that the rate of out-migration from Iraqw'ar Da/aw did not gather its full momentum until roughly the 1950s and 1960s, when it appears to have kept a steady pace with population growth. In the 1930s and 40s, and again from the 1970s, the population of Iraqw'ar Da/aw increased, indicating that the rate of out-migration from the area stagnated some time after 1970. In fact three approximate stages of demographic change since the 1930s may be identified: first, c. 1930–50, a period of population increase; second, c. 1950–70, stagnation due to massive out-migration; and third, c. 1970–90, a second phase of population increase.

A few further comments on the figures in Table 4.1 are necessary. The higher density in Kainam Ward reflects its relative proximity to Mbulu Town. The southern part of Murray Ward is the most remote and least populated part of Iraqw'ar Da/aw. The figures for 1967 and for 1934 are based on dot maps, where dots within present-day Iraqw'ar Da/aw have been counted, while the other figures are estimates of the total population of Iraqw'ar Da/aw (Kainam and Murray Wards). The figures derived from the dot maps (1967 and 1934) are comparatively lower than those of the published census data. It is, therefore, unlikely that the 1967 figure marks a decline in population in that decade; more likely, it reflects the different methods and data used. The figure for 1934 is even more conspicuous. As Winter and Molyneaux (1963) pointed out, the first house-to-house census in 1948 demonstrated that previous population figures for Tanganyika as a whole had been underestimated by approximately a quarter, so we may thus suspect that the 1934 estimate errs on the low side, but in any case it seems that the Iraqw'ar Da/aw population increased during the 1930s and 1940s. The estimated figures for population distribution, used by Gillman in his population map of Tanganyika Territory, were compiled at the Mbulu District office, and reflect

Table 4.1 *Population data for Iraqw'ar Da/aw*

Year	No. inhabitants	Density (persons per square km)
1988	**19,048** [1]	Kainam: **112**, Murray: **66** [2]
1978		Kainam: **101**, Murray: **58** [2]
1967	**16,000** [3]	80
1957	**17,800** [1]	89
c. 1955		**77** (part of Murray) [4]
1948	**17,529** [1]	88
1947		c. **48-86** [5]
1934	**c. 10,000** [6]	50

Notes to Table 4.1: Bold figures indicate data extracted from the literature. Blanks are due to the inadequacies of the secondary sources

1 From Snyder 1996, based on census data.
2 Meindertsma and Kessler 1997: 26.
3 Schultz 1971, Map 22: Population distribution in 1967. The figure is derived from the dot map, which in turn is based on census data.
4 Based on a survey by Fosbrooke in one clan area in Murray village (1955: 53).
5 Estimates made in 1947 just before the census in 1948, over an area of c. 680 square km on the central Mbulu Plateau (including Iraqw'ar Da/aw), yielded a figure of 48 inhabitants per square km. But in the most densely populated areas there were probably c. 86 inhabitants per square km (see footnote 14) (Rowe 1947).
6 Dot map in Gillman 1936.

normal colonial procedures in estimating population figures (Gillman 1936, appendix). Gillman regarded the 1934 population figures as reasonably reliable on the grounds that the assessment of administrative information was of sufficient quality and that he had personal knowledge of the area (1936, inset on Gillman's map and appendix). The population figures presented in Gillman (1936) for the Mbulu District as a whole were 96,500 in 1934 (i.e. the figure upon which the population map is based), and 98,800 in 1931 (based on census data; 1936: 371). The close agreement of these two figures for the mid-1930s encourages one to use Gillman's dot map with some confidence and, tentatively, to compare it with later census data for Iraqw'ar Da/aw (as in the table).

In earlier estimates of population change in Iraqw'ar Da/aw the increase during the early half of the twentieth century was not recognized. It was assumed that the size of population in Iraqw'ar Da/aw remained stable from 1890 to 1948, again as a consequence of out-migration (Winter and Molyneaux 1963). Fosbrooke makes a similar observation about the period 1850-1950:

> about 1850, the Iraqw were pushing out from Kainam [sic], either carving out forest land (Gihandu), or settling amongst the Tatog in Central Mbulu ...

> Now at present the Kainam area holds about 4,800 tax payers and is more or less full to capacity: *it is unlikely that the tribe would have expanded into outlying areas while there was still room in the cradle area*, therefore one can deduce that the population about 1850 must have been about 4,800 adult males or, allowing for the unmarried men (25%) say 3,600 families. (Fosbrooke 1955: 26; italics added)[14]

In one sense the statement given special emphasis here demonstrates Fosbrooke's belief that maximum carrying capacity was reached for Iraqw'ar Da/aw around 1850, after which territorial expansion was the most rational reaction to population growth. But considering the lack of empirical evidence for this assumption, it provides a good illustration of how much influence the siege thesis, based on the notion 'necessity is the mother of invention', has had in Iraqw historiography.

It is doubtful therefore whether the population of Iraqw'ar Da/aw in 1890 was as high as 17,000, the figure estimated by Winter and Molyneaux (1963), or 21,000 in 1850 as suggested by Fosbrooke. Instead, the population in the latter half of the nineteenth century was probably closer to 10,000 inhabitants (the population estimate of 1934 probably being an undercount, see above). Such a figure would be consistent with the conclusion that the Iraqw population probably increased during the nineteenth century and early twentieth century as a consequence of in-migration and cultural assimilation. Thus what has previously been regarded as an early expansion from Iraqw'ar Da/aw into the surrounding parts of Mbulu Highlands may be more aptly described as a result of in-migration from the surrounding lowlands.

Intensification and expansion: discussion

From the evidence of photographs and other historical sources it has been argued that agriculture in Iraqw'ar Da/aw was intensified and that intensively cultivated areas expanded during the last hundred years. Cultivation of crops has also become gradually more important relative to livestock production in the combined agro-pastoral production system. Although this process of agricultural intensification cannot be measured precisely, useful information can be generated from historical-geographical observations.

Population growth has clearly been linked to agricultural intensification in Iraqw'ar Da/aw. But as in-migration and incorporation of people from other ethnic groups contributed to expanding the labour force during the nineteenth century, endogenous population growth cannot be regarded as an independent driving force in Iraqw settlement history. Furthermore, the expansion that took place before the consolidation of Iraqw'ar Da/aw bears much similarity to the later settlement expansion in the twentieth century (Snyder 1993: 15-18, Lawi 1999: 27-36). What differed were mainly the speed and magnitude of the process and the relation to colonial rule. Hence, agricultural intensification occurred while the cultivated areas were expanding.

Lawi comes to a similar conclusion in the context of the transition from wooden to iron implements in the early twentieth century:

> The expanded cropping described by Jaeger is probably as much an outcome of this introduction of a new agricultural technology as it is of increased human population; the latter bringing into the equation both increased labour power and food needs. From these clues one may infer that in the precolonial times land under crop cultivation gradually expanded with the introduction of iron technology and as a result of population increase. *It is however evident that this expansion went hand in hand with intensification.* (Lawi 1999: 24-5; italics added)

Some decades after the supposed Maasai siege had been resolved during the height of British colonial administration in Mbulu, the Iraqw were again thought of as being more or less under siege, in the eyes of the British, but this time by pressure from the tsetse fly. As a consequence, vast areas of bush were cleared by means of forced labour. The colonial argument motivating this indiscriminate clearing was that tsetse was invading the district from the west, resulting in settlement congestion and land degradation. Lawi (1999: 114–71) has, however, demonstrated how this observation was part of a degradation narrative, that is mainly a hypothetical misapprehension, rather than a real threat to the environmental sustainability of the area. Nevertheless, in British colonial reports Iraqw'ar Da/aw was described as crowded and congested, when in fact it was not (see discussion above). It should also be noted that the Iraqw'ar Da/aw area on Gillman's map (see Figure 4.11) is significantly more sparsely populated than its immediate surroundings to the west and north, which suggests that the supposed congestion applied more to areas outside Iraqw'ar Da/aw.

Clearly, the lack of empirical data has given ample room for loose explanations of contemporary and historical processes in Iraqw society, largely guided by theoretical assumptions. For example, the colonial concern with high population pressure has resulted in contradictory interpretations, both Malthusian (degradation) and Boserupian (intensification). The supposed historical siege by the Maasai described in the colonial literature was assumed to have forced the Iraqw to develop sustainable intensive farming practices, and the early and middle twentieth-century siege and congestion caused by tsetse were thought to result in degradation of the environment. Hence, the Malthusian scenario was applied to current developments and prognostications about the future, and the Boserupian argument to the past. Further, in the Iraqw case, these contradictory theories may indeed have reinforced each other, making the arguments even more doubtful and inconsistent. In particular, the assumption of a nineteenth-century Maasai siege may have been influenced by the colonial descriptions of the central parts of Iraqw territory as an overpopulated and crowded enclave.

It is also important to note that much of the European experience of Iraqw settlement expansion occurred at a moment when it was quickly gathering momentum along with population growth and political stability

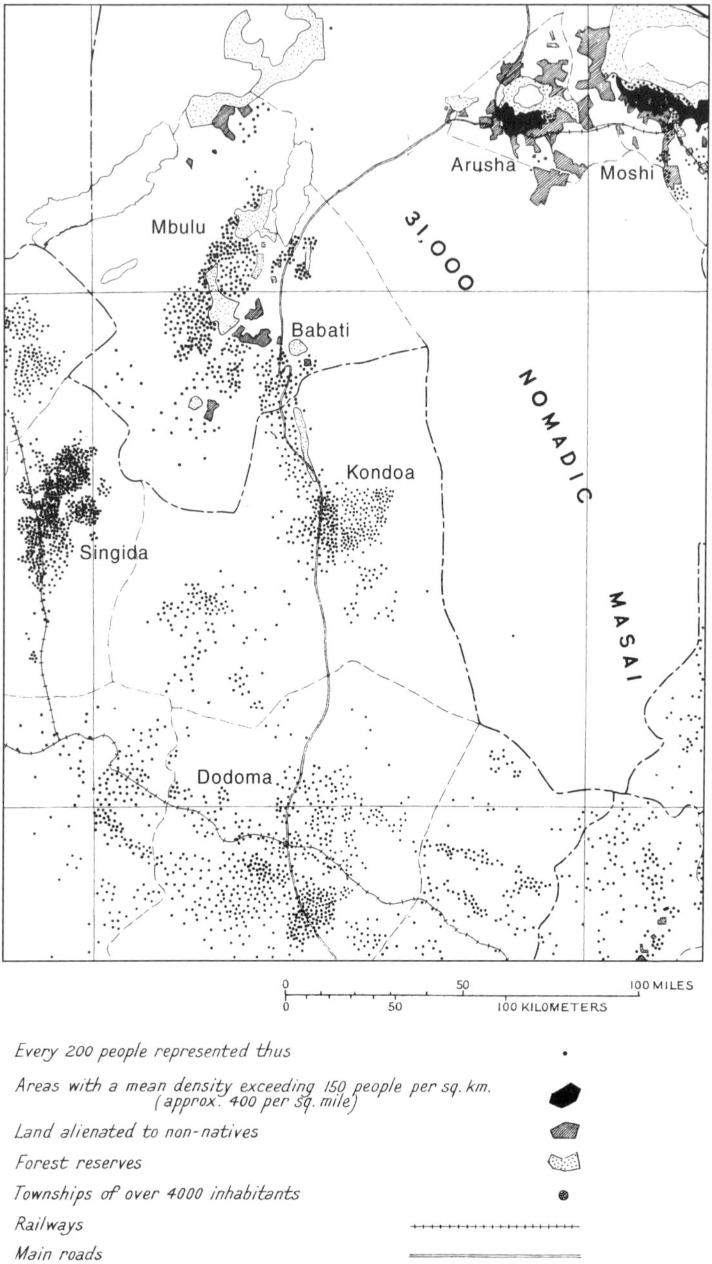

Figure 4.11 *Population map of Tanganyika Territory 1934 (section reproduced from Gillman 1936; place names added)*

during the 1920s and afterwards. In a way similar to their concern with soil erosion, colonial officials misinterpreted the expansion as a new process. The Iraqw, no longer hemmed in by external forces, were imagined as moving out from their crowded homeland, like water rushing from a breached dam. The weakening of pastoral peoples by the *Rinderpest*, followed by the European pacification of the same people and then the tsetse clearings, resulted in a rapid expansion of Iraqw territory and out-migration from Iraqw'ar Da/aw during the twentieth century. But, as demonstrated above, this was not a direct result of population pressure pushing people out when the chance occurred. The history of settlement expansion and migration among the Iraqw in the Mbulu Highlands is probably both much older and more complex than that. Hence, Schultz's observation (1971: 98) that in the 1960s the Iraqw expanded to new areas, while plenty of land was still available in the older settled areas, seems to hold as a general rule in the long history of Iraqw expansion.

How then may one explain the development of intensive agriculture in this hilly highland area, if not by population pressure? Obviously, the answer to this question is complex, but the discussion above has touched on a few contributory factors. It has been argued that intensive farming practices were adopted early on in the Mbulu Highlands, and hence that intensification has not followed an evolutionary path. Instead, it has occurred alongside the expansion of cultivation. So what kind of intensification is this? First, intensification, as conceptualized here, is the process whereby output or input (of capital or labour) increases over a defined area and time unit. It is therefore necessary to define a fixed space in which to assess intensification. Iraqw'ar Da/aw is one such space but, as has been shown, the boundaries of this space have changed and been defined differently over time.

Alternatively, one might consider the changes which have taken place in a delimited part of the historically settled core area of Iraqw'ar Da/aw. In such an area increased cropping frequency over time would obviously be part of an intensification process, but there was no real pressure to force such a development to happen. Nevertheless, such an intensification occurred, parallel with an increasing density of population, at least during the twentieth century. Furthermore, some of the characteristic intensive farming practices associated with Iraqw'ar Da/aw agriculture (such as stalling and manuring) may indeed have been practised in the area for a much longer period, but the frequency and extent to which these were put to use have probably changed through time. It is, however, difficult to find any clear shifts in farming technology which would have affected agricultural intensity and expansion, apart from perhaps the wider use of iron hoes. Thus intensification in Iraqw'ar Da/aw must be seen as an incremental process containing its own driving forces, such as the needs for food security, risk avoidance, planning and control as well as socio-political incentives, rather than being motivated by external force and internal pressure.

However, some local and regional geographical conditions that most likely have contributed to the intensification and expansion of Iraqw'ar Da/aw agriculture during the nineteenth and twentieth centuries can be identified. First, the physical geography and climate in the area were probably well-suited to an early adoption of intensive farming practices (such as stalling of cattle, permanent cropping and manuring). In particular, the abundance of dry-season grazing and possibilities for cultivation round the year, owing to the dissected highland topography, surely influenced the organization and development of agricultural practices. In fact, there may not have been a labour-efficiency threshold holding back intensification in Iraqw'ar Da/aw; instead, investment in soil conservation may have increased labour productivity. Second, the fact that people of various ethnic identities were assimilated into Iraqw society and culture points to the importance of social and political cohesive forces in mobilizing labour. Although precolonial Iraqw society shows few signs of social and economic differentiation, the power and authority of ritual leaders may have contributed to this process. Or, in Thornton's words: 'Irqwar Da'aw absorbed large numbers of peoples of diverse ethnic origin who accepted the leadership and supernatural protection of the *qwaslare* [ritual specialists]' (1980: 192). The immigration into Iraqw'ar Da/aw during these last two centuries may also be taken as an indication of a secure and hospitable socio-political and economic environment. Last, the physical and economic geography of the region stimulated trade and exchange between highland and lowland regions, providing an additional economic reason for the concentration of Iraqw settlement in the Mbulu Highlands, which would have existed even at low population densities. The processes encouraging in-migration, cultural integration and settlement concentration have all contributed to the co-evolution of Iraqw society and intensive agriculture in the Mbulu Highlands.

It has been demonstrated that the theory of population pressure, supposedly resulting from external coercive pressures on the Iraqw, cannot be sustained as a major driving force for precolonial agricultural intensification in Iraqw'ar Da/aw. Instead, this siege hypothesis has been shown to be an ideological construct which is a manifestation of population pressure theory rather than a conclusion based on historical investigation. The theoretical chain of causation in population pressure theory is: first stage, limits for expansion, resulting in the second stage, mounting population pressure, followed by the third stage, agricultural intensification. However, previous understanding of Iraqw precolonial intensification has, for lack of empirical evidence, been based on an observation of the last stage only. The preceding two stages have thus been merely assumed. However, it looks as if the development of intensive farming practices in the Mbulu Highlands (the third stage) to some extent actually stimulated population growth (the second stage) in the area. A high population density may thus reflect agricultural intensification, rather than intensification reflecting population pressure. Consequently, the search for various driving forces for agricultural intensification may in such cases be misguided, in the sense that intensification may not be a process that needs to be driven by any distinct external or internal force.

As with the gender bias in Iraqw historiography, the siege hypothesis must also be seen as a product of colonial thought. The notion of an Iraqw homeland may be regarded as historical myth, in the same way as the colonial historical division of East Africa into discrete 'tribal' areas had little relevance to precolonial ethnic relations.[15] However, the analysis of the history and geography of a local intensive farming system in this chapter has shown that the theoretical assumptions based on population pressure theory, which have largely guided previous understandings of the intensification process in the study area, have their roots in colonial thought, and they appear to be both contradictory and misguided when assessed together with the historical source material. The simple reason for this misinterpretation of the situation is that the supposed population pressure was not real, but theoretical.

Notes

* Many thanks to Vesa-Matti Loiske for everything, Katherine A. Snyder and Yusufu Qwaray Lawi for helpful correspondence, John Sutton for comments and advice, Mats Widgren and Bettina Lissner for translation from the German, and Katarina Strömdahl for assistance in drawing. Otto Norberg at the Geographical Library, Stockholm University, and the Library and staff in Rhodes House, Oxford, are also thanked for helpful service.

1 Two recent examples are Pingali et al. (1987: 62) and Gourou (1991: 50). However, Pingali et al. virtually follow Iliffe (1979: 351), who in turn bases his account on Schultz (1971) and two British colonial reports from the 1940s (Rowe 1945 and Bell 1949, Tanzania National Archives, files 69/868/1/30 and 38504/1/81A).

2 One recent figure, based on six years of rainfall data from the eastern parts of Iraqw'ar Da/aw, gives a clear indication of this variation. The rainfall varied from 529mm in the driest year to 1,849mm in the wettest. Based on these measurements the mean annual rainfall was approximated to be 500-1,500mm, with an average of 1,028mm, for this part of the area (Meindertsma and Kessler 1997: 37).

3 The effectiveness of this practice in enhancing the nitrogen content of the manure is currently being investigated (Tengö 1999).

4 The Iraqw place their previous area of habitation in a mythical place called Ma'angwatay (Thornton 1980: 196-9). Although a number of speculations about the origins of the Iraqw have been voiced, it appears as if Ma'angwatay continues to be a mythical place-name, and the Iraqw origins are more complex.

5 Rekdal (1999: 41) mentions 150 clans and Snyder (1993: 97) suggests that the number may be as high as 200. Lawi (1999: 27) provides a lower figure of approximately 100 clans, tracing their origins to the mythical Ma'angwatay settlement (see note 4 above).

6 The meaning of such a statement is, however, unclear, since it is not known when a distinct Iraqw culture and language appeared. See also Rekdal's discussion on Iraqw ethnicity (1999: 39-44).

7 An indication of this intensity is provided by the labour invested in the systematic removal of unwanted vegetation by hand in the nearby and most frequently used grazing areas. Frequently, work parties were also organized to remove ticks manually from all cattle in a neighbourhood (Lawi 1999: 98-105).

8 Obviously, the drawings or prints discussed here may reflect in various ways the ideals and perceptions of Baumann, Jaeger and the other German professionals involved in producing them. Nevertheless, they comprise one of the few sources available that

provide glimpses into the landscape history of the area. Baumann's use of prints and drawings of landscapes may have had a practical reason, that is, that characteristic landscape features would have been indistinct in black-and-white photographs (see figures in this chapter). The prints are based on either photographs or sketches by the author (Baumann, title page), suggesting that their reliability as documentary illustrations is probably good enough to support the interpretations made here. Captions in quotation marks are translated from the original German.

9 Hartley's purpose was to illustrate the 'indigenous system of soil protection' in the area, which he considered to be based on labour-intensive practices such as terracing, storm drains, deep tilling with digging poles, contour ridging, intercropping and green manuring. As a result, these pictures do not tell us much about the overall land-use intensity in Iraqw'ar Da/aw, and should be used only as examples of the kind of intensive agricultural practices performed by the Iraqw at the time.

10 However, Baumann refers to the 'Iraqw village area' ('*Dorfgebiet von Iraku*'), but what this represents is uncertain. It may have included the area depicted above, and thus most of the area settled by the Iraqw, or it could be more synonymous with Iraqw'ar Da/aw.

11 The figures of percentages of cultivated land mentioned by Jaeger must be treated as rough eye-estimates. They most probably mean cultivated areas, including fallow field, in relation to grazing land.

12 This figure has been reached by tracing the extent of Iraqw'ar Da/aw (based on various oral and written sources) on the 1 : 50,000 topographical map of the area.

13 Although the figures taken from Jaeger's map should be regarded as rough estimates, the trend is probably correct.

14 Fosbrooke states that according to census figures one adult male represents 4.45 people, suggesting a total population of more than 21,000 people in the mid-1950s (1955: 53). One explanation for this high figure could be that Kainam in this case covers a larger area than Iraqw'ar Da/aw, but in any case there is too much uncertainty attached to this figure for it to be reliable.

15 See for example Niemeijer 1996 for a critical review of recent perceptions of African agricultural history.

Five

Institutionalized Exchange as a Driving Force in Intensive Agriculture

An Iraqw Case Study

VESA-MATTI LOISKE

The history of Iraqw intensive agriculture in the Mbulu Highlands in Tanzania is (as Börjeson explains in Chapter 4) burdened with much ambiguity. Most interpretations of Iraqw history are based on data from early travellers, colonial and postcolonial documentation of various kinds and recorded oral histories. By adding data on the social system collected at grassroot level, additional and important new insights may be gained. This chapter attempts to recall history by allowing a family to tell the story of how its members network and combine social and natural resources to facilitate the family's livelihood.

Phenomena that exist parallel to the modern sector and the market economy, as do Iraqw networks, are often seen as traditional. This does not mean that the networks have existed unchanged over time, or that they can be used as historical data *per se*. The networks constantly change and adjust to new conditions to secure access to natural resources to improve the security of livelihood. People have socially constructed viable networks to survive in the multifaceted environments created by the Great Rift. The strategy of networking has been one of the pillars on which the intensive agricultural system of the Iraqw has rested. The historical lesson learnt from the grassroot perspective complements other interpretations by revealing how reproduction is constructed at a local level and how the local level is connected to the wider community. In the first part of this chapter a family history is sketched which provides the basis of the historical journey. The second section details the findings and suggests other interpretations of the history of Iraqw intensive agriculture. The empirical studies on which the story of Slaa is based is explained in a note at the end.

Slaa Margwe and his two wives

Slaa Margwe, who is in his 60s, lives near Lake Basuto in Hanang District some 60km southwest of Mama Issara (Iraqw'ar Da/aw), that is in the Iraqw expansion area. He spends most of his time here, together with his first wife Tluway. He also has a second wife, Mary, who lives some 5km from the first wife's house. Unfortunately, Slaa and Tluway failed to have children, which was one of the reasons why Slaa married Mary, by whom he has seven children. The first child, a son, is already married and has a house of his own not far from Slaa and Tluway. The second child is a girl, who is in secondary school in Mbulu town. The rest of the children are still in primary school, except a boy who is still under school age.

By having two wives, Slaa has access to two farms in different ecological zones. Tluway has most of the extended-family cattle at her house. The house is located in *miombo* (*Brachystegia*) woodland, where the trees provide useful shade and where grazing is abundant in the glades. In years when the ritual leader has predicted generous rains, Tluway plants some sorghum or millet close to the house, but in most years the area is too dry for cultivation.

The homesteads in the neighbourhood are scattered and well hidden in the landscape, and it is difficult for outsiders to imagine that people live here. Most of the neighbours are Datoga who, like Slaa and Tluway, use the area mainly for grazing. Many goats roam round the courtyard, and Mary's youngest boy tries to keep the herd together. Tluway says, 'The goats are the coins' and the main source of income, while the cattle 'are the notes that have to be saved for the future'. Tluway has become an expert on cattle rearing and has a reputation in the area of having veterinary skill.

She is also knowledgeable about cattle breeding and keeps track of various bulls in the area, so that she can maximize the output from the cattle herd. She has inherited the skill of a ritual leader from her mother's clan. She claims to be 'weaker' than other female ritual leaders and works in that capacity only in the neighbourhood. Slaa remarks that she is too humble and corrects her modesty by saying that she is very much in demand also by faraway families.

The other wife, Mary, is younger than Tluway and lives in a wetter area, closer to Lake Basuto, where large areas are under wheat cultivated by the National Food and Agricultural Corporation (NAFCO), in which Mary's brother holds a position. It was through her brother that she and Slaa got the farm. Mary is a Christian and is involved in the Lutheran congregation, and her husband keeps to the traditional beliefs of the Iraqw. This does not bother Mary much, because she is confident that Slaa will eventually also become a Christian, as he has allowed all the children to be baptised.

Mary has specialized in farming. The area is suitable for cultivating maize and beans, the staple food items in the area. She also has access to a small portion of land in a *mbuga* (wetland), where she cultivates vegetables and sugarcane. She keeps a few cattle for milk and some goats for meat, but says that she prefers cultivation. Slaa explains that Mary has green fingers and that she has become a skilled cultivator, although, when young, she was a bit lazy. Slaa and Mary laugh at the recollection. 'Reality comes with the children', as Mary puts it.

Mary is also involved in petty trade. She buys mudfish from the fishermen at Lake Basuto and smokes them. The fish are then sold to traders from other towns involved in the fish-retailing business. Mary says that she takes part in the fish business only part-time, as both access and demand for the fish vary, 'but with the money I can buy new clothes now and then'.

Tluway and Mary complement each other in giving the family access to both cattle and crops. They also provide Slaa with two additional clans who care and are cared for by the family. He has a multitude of contacts with his own clan. When Slaa was young and newly married to Tluway, they were both living in Mama Issara, Slaa in Hhay Qafa and Tluway in Amowa in Kwermusl village. It was very clear that Slaa, when married, would have to leave Mama Issara, together with two of his brothers, as his father's land was too small to divide among all the sons.

They did not move into completely unknown territory, however. Slaa had been looking for suitable locations for quite a while before his marriage. The particular location he chose was recommended by other Iraqw who lived close by and who also had many cattle to tend. Slaa also had several clan-mates living in the area whom he particularly trusted and who recommended the area. When Tluway saw the area for the first time she said some traditional prayers and blessed the compound in the way she had been taught by her mother. After the blessing rituals, she agreed that the place was good enough for them to start their new life.

As Slaa wanted to be a herdsman and was prepared to move to an area with abundant grazing, he and Tluway were given the important task of taking care of the 'grandfather's' cattle. His father and uncles inherited the 40 head of cattle that had been accumulated by Slaa's grandfather in Mama Issara. Since grazing there was limited, the family decided to move the cattle. The marriage between Slaa and Tluway was very timely, as the grandfather had died just a few weeks before. The cattle that he had accumulated during his lifetime had been distributed among many friends and relatives. They were now collected and became an important incentive for Slaa and Tluway to start their new lives as cattle rearers.

Having cattle as a basis for household production requires that one has good contacts with people who concentrate instead on cultivation. Some contacts were recommended by both Slaa's and Tluway's parents in connection with the wedding. Others they had to arrange themselves. The family was very dependent on exchanging goats and milk products to get cereals in the first few years. They therefore developed contract exchange with other households which had products that they needed and could not produce themselves. Maize, honey, sorghum, salt, fish, vegetables, tobacco, sugarcane, herbal medicines and other items were obtained by exchange in this way. Some of the exchanging households had access to early maize and others to late maize, and Slaa and Tluway had to find a combination of families with which to exchange in order to have supplies for the family's needs throughout the year.

Preferably, such exchanges are made with households in one's own clan. Slaa and Tluway had exchanges with one family outside their clan, and that was the family of an old friend and neighbour from Mama Issara whom they both considered to be very close. The reason why members of one's own clan are preferred is that reciprocal obligations are respected better within the clan than outside. The exchange arrangements constitute therefore a social-security network. In times of hardship, the families involved in an exchange network assist each other. This help is institutionalized, as is the network.

When Slaa married Mary, several of the items to which he and Tluway had access through the exchange networks became available within the family. This did not, however, mean that the existing exchange networks were abandoned. On the contrary, Mary's inclusion in the family meant an increase in numbers, as Mary also brought into the network exchange families from her clan. Slaa says that the greater the number of exchanges, the more secure the family is.

In talking about the various potentials of the two places that Slaa's family has, he himself came to appear idle, not himself providing any income or labour to the household. Slaa, being a man of few words, suddenly started to talk about his 'bull business'. He sells a mature bull at the monthly cattle auction and buys three or four bull calves from the income. He feeds the bull calves for some years and then sells them. He makes a good profit from the business, as meat has become a much appreciated part of the local diet, especially by those who eat at restaurants in the nearby towns.

Continuing to talk about bulls and his business, Slaa explains the whole hidden economy of the Iraqw. He says that the market is a very 'difficult relative'. Prices fluctuate, people fool each other and money is inflated. The traditional exchange between clan members is safer. It is a long-term contract between families, forming connectedness and mutual dependence between those involved. Other clan members guarantee that the exchange is fair; when doubts occur, the clan elders will have a word to say. A bag of maize is always a bag of maize, and exchanging a goat for salt is done according to the traditional exchange rate which does not vary with time. Any Iraqw therefore prefers traditional exchange.

Slaa, Tluway and Mary have some 20 households with whom they exchange, and they think that they could have more. Slaa comments on that by saying that through this kind of exchange all Iraqw families are connected with each other. The larger the number of families connected with each other, the better for the Iraqw. If one part of the Iraqw world, say Mama Issara, faces food shortages for some reason, all the Iraqw will help that part out. 'Even if my exchange never reaches Mama Issara, some of those I have exchange with will ... they will be my extended arm.'

Ritual networking is also important. The exchange and ritual networks partly coincide and are partly different. While the exchange of food items diminishes with distance and does not extend beyond 30km, the ritual contacts always continue with the core area, Mama Issara. The reason for this is that Mama Issara forms the cultural centre of the Iraqw. It is difficult for outsiders to understand fully the importance of the role played by Mama Issara.

Slaa still visits his clan relatives in Mama Issara, where important clan issues are discussed. Slaa belongs to the group of 'wise men' (elders) where he lives, which gives him a fairly high status in the clan. One of his uncles in Mama Issara is the most important person in the clan. Slaa visits him several times a year to settle various clan issues. Among the issues discussed are weddings, funerals, blessings of the soil, harvest feasts, the clan land and the traditional penalties that sometimes have to be paid by the clan. Slaa also takes care of his wives' clan issues when he goes to Mama Issara. The women are not expected to make the long and dangerous walk to Mama Issara, and they therefore visit clan relatives living nearby more often. Slaa foresees the time when he will be too old to walk to Mama Issara, but thinks that his eldest son is still too young to perform such an important task.

Winding up, Slaa remarks that the Iraqw have managed to compete successfully with their neighbours for two reasons: diversity in production and close networking between households. 'Being single is to be fragile.' Therefore the Iraqw are agro-pastoral and have consciously chosen diversification as a strategy. In the highly variable environment of the Iraqw area, diversification means connectedness. The example of Slaa and his wives gives us some new insights that will now be discussed.

Networking

As illustrated by the case of Slaa, various forms of networking are important in Iraqw society. Three specific types of networking can be identified: networking involving exchange of agricultural produce, networking with cattle and thirdly ritual networking, the families involved being often of the same clan. These three forms of networking often overlap and intertwine. Networking forms the main livelihood strategy among Iraqw families and, furthermore, consolidates the larger community to which individual Iraqw families are connected. Clans also connect to each other through these networks as husband and wife or wives are often from different Iraqw clans. Networking can therefore be seen as the cement that, at least partly, holds Iraqw society together and keeps it going.

The physical characteristics of the area where the Iraqw live, with varying elevations, soils and hydrological conditions within short distances, provide a diversity of ecological niches with varying agricultural seasonality and cropping opportunities. Networking is therefore essential for access to resources not available in the immediate home area. The networks provide agricultural produce, salt, herbs and other products. It also gives an opportunity to gain access to products in off-seasons.

Iraqw cattle owners, as is common among various ethnic groups in Africa, pursue an opportunistic grazing strategy by distributing cattle over wide areas and to many families within their networks. The strategy is designed to serve three main ends: leaving sufficient cattle at the homestead to provide for milk and manure; distribution of stock to families in various areas within the network, to reduce the effect of theft, diseases or droughts and to diminish labour constraints; and finally to concentrate family cattle ('grandfather's cattle') in areas where grazing is abundant and where cattle thrive best, as exemplified by Slaa and Tluway. Cattle networking also provides a wealth distribution system, as many resource-weak Iraqw need to complement their herds or, if they do not own cattle, borrow them from more well-off farmers.

Ritual networking handles issues related to the clan. The male clan elders take care of the wellbeing of their clan-mates, on both their own and their wife's or wives' side. There are numerous ceremonies to be organized in the year, such as burials, weddings, harvest feasts and traditional punishments within the clan. The clan elders are responsible for the proper implementation of these ceremonies and for matters concerning marriages and inheritances. One aspect of the ritual networking is the search for remedies for diseases either at individual or regional level. These remedies are often found in a dialogue with ritual leaders who prescribe the correct medicine. The ritual leaders often come from other clans and have the role of taking care of Iraqw society as a whole. They form another part of the cement which keeps Iraqw society together. Female ritual leaders have a more domestic role, as exemplified by Tluway, and often make home visits to families with physical diseases or mental disorder. They can be considered as barometers of social health at the local level, and the people who sound the alarm when communal illnesses occur. They also bless homesteads when disease threatens livestock, humans or crops.

Networking provides social security as the cooperating partners help each other out in times of strife. If one of the networking families refuses to share its resources with families in need, it may be punished according to clan rules which may involve fines, such as a certain amount of beer or, in severe cases, livestock. The most important function of these networks is to provide a framework for strategies of risk aversion.

These strategies vary according to location. For instance in Mama Issara, where Slaa and Tluway were born, there is a favourable water balance which supports many wild plants used as vegetables as well as a variety of crops. It is possible to obtain two maize harvests each year. The strategy here is to increase agro-diversity *in situ*. Most families grow a variety of crops and keep enough cattle for milk and manure. Moreover, for security during times of drought or excessive rains, families in Mama Issara involve themselves in exchange networks with families outside the area. Outside Mama Issara physical conditions vary and the strategy of *in situ* diversification is difficult or impossible. In the most remote places, agriculture is restricted owing to lack of rain, and cattle rearing predominates. In very specialized ecological niches, sugarcane, garlic, onions, tomatoes, early maize, tobacco, etc can be grown. Many Iraqw outside Mama Issara have had to specialize in specific crops adjusted to the local conditions. To achieve the diversity desired, the Iraqw have developed an institutionalized network, *inaoha*, literally meaning exchange. If only one or a few crops are cultivated, these can be exchanged for crops produced in other areas, to increase diversity. Exchange is agreed in advance between households. There is less exchange when the distance exceeds 30km because the transport cost will be too high. Outside Mama Issara, the Iraqw tend to use *inaoha* to achieve the desired diversification.

The average household in the expansion area cultivates one crop a year, the maize harvest being in July. In Mama Issara, by contrast, the major maize harvest is in January, which means that fresh maize is available at different times in the district. Gaining access to fresh maize is one particular reason for exchange arrangements between Mama Issara and the expansion areas, thus providing relief to families when their own store is empty. In this way diversification can be achieved round the year.

Small livestock, most often goats and hens, may be exchanged, as in the case of the cereals above. With cattle, several institutions are maintained, the most common being a borrowing system in which an owner lends cattle to a poor man lacking stock of his own. This serves as both a labour-saving device and a means of distributing wealth. It is common among the Iraqw, as few individuals actually own cattle, whereas through this borrowing system most households have some access to milk and manure. Another common institution is that of 'grandfather's cattle', in which a relative living far out in the dry expansion area takes care of clan cattle owned by several people, for example, brothers. As grazing is limited in Mama Issara, such surplus cattle have to be kept in the expansion area.

Specialization in certain products in a particular area has encouraged geographical divisions of labour. As in the case of the Marakwet irrigation farmers (see Chapter 2), the varying preconditions for agriculture and grazing have resulted in specialization and exchange in Iraqw. At the most obvious level, there is the division between pastoralists and cultivators who are mutually dependent on each other for survival; then there are cultivators who specialize in different crops. The integration of pastoralism and cultivation in the same household, as occurs in Mama Issara, or by exchange between households in different areas, are ways of integrating production in an efficient manner throughout the Iraqw area and of ensuring increased security and better living conditions for families.

All such networks and strategies result from a dynamic social structure still allowing for changes when need be. In the case of Slaa's second wife, Mary, it was the Christian congregation that became attached to the family network, even though Slaa is a non-Christian. Another example is how Mary through her brother managed to involve NAFCO (as noted above) into the family network and gained access to farmland which could be used by her and Slaa.

These two examples suggest that the Iraqw social system is adaptable to new circumstances. Another pointer in the same direction was encountered when clan genealogies were drawn. The informants often had to consult other clan members to obtain a consensus on their forefathers: who and how many they were, and the order of succession. It sometimes took several weeks for the clan to reach this consensus. The history of the clan affects issues of ownership, of both land and cattle, of marriages and divorces, of traditional punishments, etc, and it is considered beneficial to be flexible when resolving such issues within the clan.

Slaa, his wives and Iraqw history

This case of Slaa and his wives is offered as a complementary study to the historical account of the Iraqw given by Börjeson in Chapter 4. The most persistent theme in Iraqw historiography has been the siege hypothesis (as explained by Börjeson). By this view, Mama Issara is said to have been isolated for a long period of time. However, the networking strategy of the Iraqw as presented above testifies that exchange networks remain important for survival in this area. Nothing in Iraqw history, if one rejects the siege hypothesis, suggests that there have been survival strategies other than networking since early settlement in Mama Issara. Lower population densities in the past, and a less versatile level of technology resulting in lower crop yields, meant that provisioning would have been more difficult to achieve than nowadays. This suggests that the incentives for exchange networking would have been even stronger than now. These factors may also have strengthened the incentives for the spatial division of labour, since intensification of agriculture *in situ* requires more labour. The harvests would have been especially dependent on the local soil conditions and seasonality before the recent introduction of new crops and high-yielding varieties. Instead, farmers in the past had fewer options and had to specialize either in crop production or pastoralism. This might be called the network hypothesis. A further factor in support of this hypothesis is the need of a cattle population for a larger gene pool than could be maintained within the small Mama Issara area. Too small a cattle population and local gene pool remains a problem even now, and would have been more severe with a lower human population and lower cattle densities. Furthermore, salt and iron for hoes and other tools were needed by the Mama Issara agricultural community and must have been obtained by exchange with people outside the immediate area, as no indication of iron making or salt extraction has been recorded within.

Another feature that becomes clear in the Slaa case is the dynamic character of Iraqw networking and life. When new opportunities come forward or existing ones change, the networks are modified and adjusted to the new situation. This illustrates how intensive agriculture, specialization in agriculture and cattle rearing form part of a mechanism by which the wider Iraqw society manages to adapt to new circumstances. Networking makes Iraqw society flexible enough to incorporate new opportunities and to revive defunct ones. This helps to explain the persistence of intensive agriculture in Mama Issara. Iraqw society is sufficiently flexible to incorporate new features, but at the same time sufficiently conservative to keep up the traditions and social homogeneity. Ritual networking is instrumental in the reproduction of 'Iraqwness' and in assimilating individuals from other ethnic groups. This faculty for incorporating others was one of the pillars of success of Iraqw society during the twentieth century.

The social system of the Iraqw is thus both sufficiently dynamic and conservative to preserve the Iraqw identity even as the world around is changing rapidly. The social cohesion created through intensive and geographically shared networking assures the Iraqw as a group a comparative advantage in the region, one which means culturally that old is integrated with new, and that older features are adapted and innovations incorporated into the continuing tradition. The survival of intensive agriculture in Mama Issara has to be understood in this context.

Note

The fieldwork that provided the empirical background to this paper was conducted between 1991 and 1997. The village of Giting in the expansion area, some 120km south of Mama Issara, was the focus of the first study in 1991-5. The study examined why the villagers failed to take care of their land in a sustainable way when it was so obvious that they had the knowledge of good land husbandry, related as they were to the Mama Issara Iraqw. In the second period (1995-7) research was focused on the social, economic and ritual links between Iraqw living in Mama Issara and those living in the expansion areas, with the aim of understanding how and why Mama Issara still exists. It was a tracer study, in which 60 households, 20 in Mama Issara, 20 in a zone 20-40km south of Mama Issara, and 20 in the most remote and marginal area far away from Mama Issara, were interviewed. By constructing genealogies and networks of both husbands and wives in the 20 Mama Issara households, 20 relatives who had moved out of the area could be identified. They in turn were traced and interviewed in the same way as the Mama Issara households, and 20 of their relatives who had moved even further into the expansion area were identified, traced and interviewed. The families were visited three times (sometimes more), firstly to carry out an in-depth interview, secondly to adjust genealogies and networks, and thirdly to make final verifications. During the one and a half years that the empirical part of the study took, most Iraqw areas south of Mama Issara were visited. The detailed findings from these studies are yet to be published.

The names of the individuals, Slaa Margwe and his wives Tluway and Mary, have been altered to preserve their identity.

Six

Engaruka
The success & abandonment of an integrated irrigation system in an arid part of the Rift Valley c. fifteenth to seventeenth centuries

JOHN E.G. SUTTON

Archaeological testimony in geographical and historical perspective

Engaruka is unique, partly for what it was, but no less spectacularly for what it is, namely the visible remains of a completely abandoned but formerly expansive irrigation system with villages which flourished at the foot of the Rift wall (at 3° south), between Lakes Natron and Manyara, roughly between three and six centuries ago. In this rather arid spot in what subsequently became sparse grazing territory for Maasai, with average annual rainfall of less than 400mm – that is well below the minimum necessary for normal cultivation at this latitude and altitude (about 1,000m) – the ancient community constructed canals off the Engaruka River which descends from the Crater Highlands through a deep gorge in the escarpment, and also off lesser seasonal streams issuing from the escarpment immediately to north and south. The field system irrigated by these means can be traced along a 9km stretch of the escarpment foot and covers an area of at least 2,000ha (Figures 6.1 and 6.2). The main crop, judging from the results of excavations in the old Engaruka villages, as well as from the examples of recent and existing cultivating communities using irrigation in relatively dry terrain in the broader region, would have been sorghum.

There is nothing very unusual about irrigated agriculture, with indigenous roots in the precolonial past, along the Rift wall of northern Tanzania and northwards into Kenya, from the Manyara escarpment at 4° south, through the celebrated instances of Sonjo, Baringo and Marakwet, and on to Seger in the Pokot Hills at nearly 2° north, or again to the east of the Rift in the

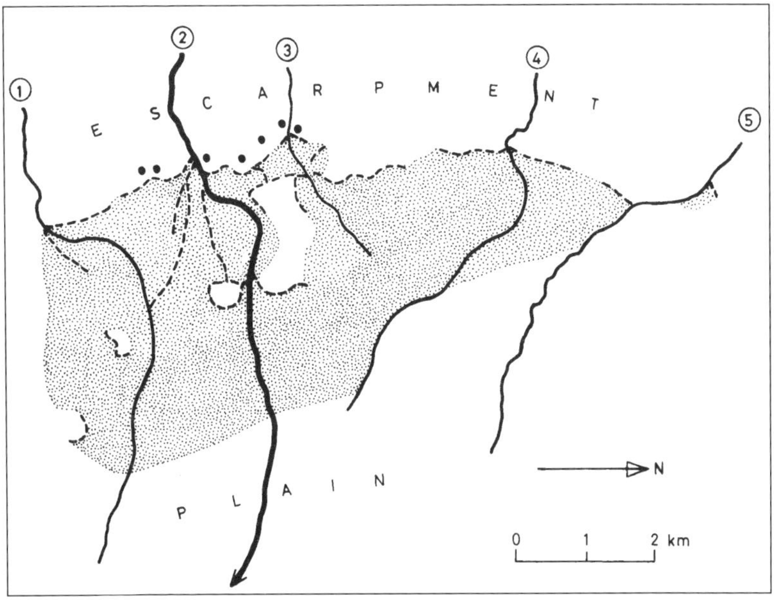

Figure 6.1 *The Engaruka ancient field system, showing the Engaruka River (2) and seasonal streams (1, 3 – 5), the ancient cultivation area (stippled), the main artery canals (broken lines) and the villages on the escarpment scree (black circles). The blank areas among the fields are hills which rise above the levels which the channelled water could reach. (Map: J.E.G. Sutton and G.M. Oteyo)*

highlands of northeastern Tanzania and the Kenya border, notably Pare, Taita, Kilimanjaro and Mount Meru (see Chapter 1 and Figure 1.1, and *Azania* 24, 1989). Among these communities numerous varieties of field systems are found, prepared for different crops and combinations and depending in greater or lesser measure on artificial irrigation of the hill-furrow sort, that is by constructing small gravity-fed canals off springs or mountain streams. But since these present and recent fields are obliged for obvious reasons (the water sources and the basic hydraulics of gravity-fed furrows) to use much the same irrigable land as the older ones, the latter are barely recognizable as such on the landscape. Or rather, where there is a strong suggestion of continuity of settlement and of cultivation dependent on irrigation over a long period, one may, as a historian, have to be content with regarding the ancient and the existing fields as all one. At best therefore, in these favoured areas of concentrated agricultural settlement, one rarely gains anything more than an impression of the area of ancient fields and of the cultivation and irrigation system which operated in the past.

At Engaruka, in contrast, the earlier field divisions and the irrigation works running through them are fossilized on the ground and are traceable, as mentioned, over 2,000ha. The reason for this survival over so large an area

Figure 6.2 *The central field area, looking west towards the escarpment base and the Engaruka gorge (central). (All photographs: J.E.G. Sutton)*

is twofold. First, although Engaruka, with its seven large nucleated villages associated with this extensive field system, must have been a very successful settlement for a period of two or three centuries, from at least the fifteenth till the seventeenth or possibly the eighteenth, eventually it failed and was completely deserted, so that the fields, canals, homesteads, etc survive on the ground in a fairly undamaged state. This unusual archaeological visibility is equally dependent on the second factor, the use of stone. The Engaruka fields were divided, retained and levelled for irrigation with stonework; the canals and the feeder furrows were similarly lined (and in places carried as causeways on stone embankments). Moreover, in the villages (black circles on Figure 6.1), situated on the steep escarpment scree above the fields and highest canals, the individual homesteads were built on levelled platforms revetted with drystone walling at front and rear (although the houses themselves were of wood and thatch). This intensive use of stone was not a matter of choice, even if it may have become in a way a cultural hallmark of this community, and have afforded their villages and field and irrigation system an appearance of permanence. Essentially it was necessitated by the sheer frequency of loose stone on the escarpment scree and all over the irrigable area below. Every minor field plot needed to be cleared of stones, and intensive cultivation and irrigation would have been constantly bringing more to the surface. In some parts of the fields the excess of these stones, over those

Figure 6.3 *Terraced field divisions in the south fields (with escarpment base in background)*

Figure 6.4 *Stone clearance cairn, now bearing a termite heap (following decay of a tree). The break in the drystone casing reveals the rubble interior*

which could be employed in field divisions, terraces and furrow lines (Figure 6.3), was simply piled into square or angular cairns, which were revetted with near vertical drystone casings of larger boulders in order to waste as little land as possible (Figure 6.4). Others were used for building walled enclosures among the fields; these were presumably for penning cattle, which would have been stall-fed so that manure could be available as fertilizer in the fields around.

The wide extent and the sheer density of the stone field divisions with the double-lined canals and feeder furrows running through the grid – as well as the clearance cairns, the stone enclosures and the large concentrated villages, comprising altogether several hundred homesteads situated immediately above the top canals – lend Engaruka an impression of having been an unusually intensive agricultural system. This impression is partly due to the propensity of observers to be enthralled by an unusual archaeological record with such visible stone remains (the 'tyranny of the monuments', in Ian Farrington's phrase: see *Azania* 24). But it cannot be dismissed as entirely a case of false or exaggerated appearance. In this arid and uninviting terrain, with such low rainfall and high evapo-transpiration, no cultivators, however hardy and diligent, can have contemplated settlement without being prepared to irrigate artificially. Although they would – like the new community of Engaruka which has settled a short distance downstream over the last 100 years – have made the best use of the rain when it came, and might, in certain good years while the population was still quite small, have managed with few and simple irrigation arrangements, there would have been from the beginning a definite dependence on irrigation, or at least a readiness to fall back on artificial watering and to extend it in any particular year.

In this respect Engaruka contrasts with a number of the existing and recent irrigating communities in the broader region. Some of these, notably on the southern slopes of Kilimanjaro and Meru, enjoy quite heavy rainfall in most years, so that irrigation may not seem at first sight to be essential for cultivation. It has, however, become essential to those communities and an integral element in their agricultural systems as they have successfully evolved, the population density having increased beyond the limit which can be sustained from rain-fed fields alone. Furrow irrigation has therefore been developed in practicable valleys and basins in order to extend the planting season and to provide flexibility and crop combinations round the year. In certain cases there has also been a commercial factor, in particular growing coffee for export on Kilimanjaro and the North Pare hills in the twentieth century, as well as market vegetables. In the nineteenth century, moreover, the irrigated basins of such places as Baringo in the Rift, Taveta on the road east of Kilimanjaro, and the villages at the east foot of South Pare acted almost as oases in supplying trading caravans before and after crossing the drier plains. And at the other end of the spectrum of furrow maintenance one must not overlook the usefulness of ensuring a ready supply of running water close to homesteads (and stalled livestock), especially in densely populated areas such as the slopes of Kilimanjaro.

The Engaruka irrigation system: Water sources and furrow technology

Engaruka belongs in a different category, since the rainfall here is definitely inadequate for attempting cultivation without devices for supplementing it through artificial watering. This is facilitated by the physical environment of Engaruka. Above the sheer Rift escarpment stand the Crater Highlands (Figure 6.8) which catch twice or thrice the rainfall of the Rift floor. Once the montane aquifers are saturated, the excess water drains down the escarpment through deep gorges, of which that of the Engaruka River is the biggest on this stretch. Where this river and likewise lesser and seasonal streams to its north and south emerge from its gorge at the escarpment foot, outwash fans have accumulated, with the result that the river enters the plain at a commanding level, ideal for irrigating land either side and as far into the plain as the volume of water can be stretched.

The old Engaruka community effected this by leading canals off both the main river and the seasonal streams at the highest practicable points in the gorges, that is at the vertices of the outwashes, and carrying them along the rocky escarpment base as close to the horizontal as possible (Figure 6.5). The actual take-off works on the sides of the stream-beds have of course not survived the spates since the time of their abandonment, but one must imagine improvised structures of stone and trash (akin to those of existing irrigation systems in Marakwet and Sonjo) which would have required annual maintenance if not rebuilding. The positions of these take-offs can be estimated fairly accurately by tracing the visible stretches of the artery canals back to source. The stone linings and embanking of these canals are preserved, quite spectacularly in some stretches, along the base of the escarpment scree, around the small hills on the edge of the plain and through the whole area of ancient field remains, as shown on the plan (Figure 6.1).

These artery canals, of which the biggest are 1–2m wide and traceable up to 3km from take-off, with major branches at various points, were carefully engineered to hold the highest practicable level over this steep and rough terrain, being secured with stones packed with earth and with gravel embanking for extra support. In certain stretches canals were raised as much as 3m on embankments, in order to maintain desired levels across small valleys or gullies and thus to allow irrigation of substantially wider areas of the plain. Some of these embanked stretches (Figure 6.6) appear to belong to a programme of relaying large areas of the field system at a relatively late stage. They suggest the need to increase the efficiency or intensity of the system, and to respond to stresses being felt once the field area had attained the maximum extent which could be reached by gravity-fed channelled water. Similarly, the highest escarpment furrows, holding as commanding a level as was feasible, must belong to a relatively evolved stage of the system, when, as a result of its early success, its population had increased to a size requiring that all possible land be brought under cultivation and regularly watered, both on the stony escarpment and in as wide an area of the plain as could be reached by irrigation. There are suggestions here of strains on the resources, with perhaps fertility of the fields declining through over-use, despite manuring and rotation of the crop repertoire (pulses, say, following sorghum as the staple), and soil eroding from them through excessive watering, despite again the careful terracing and revetting with stone. While this scenario is necessarily impressionistic, there are clear signs of the denudation of the upper fields, with their divisions and terraces standing increasingly bare and stony, and of the corresponding deposition of soil on areas of lower fields, where the lesser divisions and many of the furrows are masked, or visible only in recent gully sides. On the north side some of this depth of deposition must have occurred after the abandonment of those fields, and it is difficult to determine precisely how much of the erosional process occurred during the time when old Engaruka was occupied and its fields were being watered and farmed.

Even more crucial was the behaviour of the streams descending the escarpment, their flows becoming weaker and less persistent. This hydrological decline along the Rift wall must have had an obviously adverse effect on the viability of this substantial community cultivating in so dry a place and dependent on the irrigation. The evidence for the decline is clear in the configuration of the field area, canals and villages, as illustrated in Figure 6.1. This shows considerable parts of the fields being watered by canals led off the Olemelepo River (no. 1) at the south end and from the Makuyuni (no. 4) to the north (and even the Loljoro, no. 5, beyond that). All these streams are now seasonal, the Olemelepo and Loljoro opening only for short periods each year after heavy rain, when irrigation water would be least needed, and failing almost as suddenly once the clouds have dispersed. The Makuyuni (no. 4) is less temperamental than these and its middle stretch is used for irrigation by farmers now, but its flow usually persists for only half the year, sometimes less, and could not have supported the extent of fields irrigated from it three and more centuries ago. The same observation applies with more force to the large area of ancient fields on the outwash of the Olemelepo (no. 1). That, like Makuyuni, had canals taken off the very opening to its gorge in order to reach as wide an area as was physically possible (Figure 6.5). Any attempt to revive those canals now would be purposeless.

This process of hydrological change is illustrated even more spectacularly at the intermediate north gorge (no. 3). This is a narrow cleft whose flow nowadays opens only in occasional years after exceptional rain on and above the escarpment. However, short canals were led off either side of this gorge, and an area of fields below was laid out with a neat grid of parallel feeder furrows taken off the arteries (Figure 6.7). Immediately above those canals are the two most northerly of the old villages. It is not necessary to conclude from these observations that the stream in this cleft formerly flowed permanently; those who occupied these two villages may, for part or much of the year, have been obliged to fetch domestic water from the permanent Engaruka River (no. 2), only 1km distant, or, more conveniently, from the great northern canal which originated high on the Engaruka's bank and was

Figure 6.5 *The retaining wall for the escarpment base canal, led northward off Olemelepo (with recent break in middle of picture)*

Figure 6.6 *The embanked causeway on the great northern canal*

constructed through the valley below this gorge. It is obvious, nevertheless, that no villages would have been built in this position, and no canals would have been constructed to supply the field grid immediately below, if the flow from this gorge had been as rare as it is now. Four or five hundred years ago it must have been relied upon to open annually, just as the Olemelepo, Makuyuni and Loljoro streams must have flowed for longer seasons than they do now.

The great northern canal which was led from the main river along the escarpment face and then, assisted on embankments in places (Figure 6.6), through the valley below this intermediate north gorge, was apparently part of the answer to the decline of this seasonal stream and of others beyond. For this canal continued through the foothills, still holding a level from which a large part of the Makuyuni basin could be reached. While its burden could not be carried on to the upper part of the Makuyuni outwash – after 3km it had lost too much height for that – nor could it reach those at the base of the intermediate north gorge, it would doubtless have ensured for a while the viability of large areas of the lower northern fields, and the sustenance of the villagers who tended them, as the annual flows of the Makuyuni and other streams on this side shortened. The lower part of the Makuyuni basin moreover could have been reached by other canals led off the main river downstream of the foothills.

This meant that for large areas of fields there were, theoretically at least, other sources of watering. Whether the system was so flexible in practice is, however, more difficult to gauge. That would have depended on various social and organizational factors (as studies of existing examples of irrigation, notably in Sonjo and Marakwet, indicate). There could have been physical constraints too, since it may be wondered, from the current volume of flow of the Engaruka River, whether that would have been sufficient to supply all the long arteries on either side in dry seasons when irrigation was most needed. At the least, a strict set of controls, with rationing or rotas of water use along each canal and branch, as obtains in existing irrigation systems, would have been essential.

These arrangements by which areas of the northern fields could, at an evolved stage of the system, be irrigated by water from other sources, have their counterpart in the southern field area. Most of the land halfway between the Engaruka and Olemelepo gorges could be reached by the canals taken off either river – or, rather, it could be reached by those led off the Olemelepo during its period of flow, and at other times of the year by water channelled from the Engaruka. This possibility does not of course hold for the outwash fan of the Olemelepo, and the clear field divisions and furrows which cover it. That (like Makuyuni outwash at the other end) stands too proud for water from the Engaruka River to be carried on to it. It was irrigated from the Olemelepo by an artery canal led northwards along the escarpment from its gorge (Figure 6.5) and by another on the south side constructed along the crest of the outwash (see Figure 6.1), notwithstanding that the present behaviour of this river suggests that the flow would have been seasonally

Figure 6.7 *Grid of feeder furrows and field divisions below the northmost village and the narrow intermediate north gorge*

restricted and in some years of no help at all for irrigating. But the middle and lower parts of the Olemelepo basin could be reached either by those canals, whenever the Olemelepo was flowing sufficiently, or by canals taken from high in the Engaruka's gorge. One of these ran southwards along the escarpment scree as far as the most southerly of the villages, another transversely through the foothills, and a third was taken off the river downstream below the hills. In the upper area of overlap from these sources there are, as much as in the northern fields, instances of embanked canals apparently superimposed over earlier field and furrow alignments, and indications of wholesale relaying of new grids. These seem to represent extensions and improvements of the engineering skill and ingenuity of the irrigation system generally. But again, it is less clear whether they indicate increased or more efficient production from the fields or, rather, a reaction to declining water supplies and productivity in a highly specialized situation, with a large population to be fed and the limits of irrigable land and technological solutions being reached.

It is necessary also to consider the volume of water in the permanent Engaruka River, and whether, with the decline of the seasonal rivers, that would have sufficed for all the long canals taken off both banks to serve the 2,000ha or more of fields, even with a rota system for watering and the fields being subjected to necessary rotations and fallowing. The experience of the present villagers, who have settled at Engaruka in the colonial and postcolonial periods and who farm by irrigation (but without stonework) in the plain below the main ruin area, is instructive. In years of poor rainfall, or at the critical growing seasons, the diversion of water into the fields through several furrows can curtail the river's length noticeably, and has on occasion provoked objections from pastoral Maasai who need to water their cattle in the sparse pastures of the Rift floor. The local community has had therefore to devise rotas for opening and closing the various furrows taken off the river at intervals, as well as to ensure a flow beyond the cultivated area into the plain.

At the least, the hydrological decline attested by the behaviour of the seasonal streams some three or four centuries ago, while old Engaruka was occupied, presumably entailed some reduction in the flow of the Engaruka River too. One might try arguing that the changes were not in the volume of water descending the escarpment but merely in its seasonal distribution, with the flows in the lesser streams becoming more violent but less persistent because of environmental changes to their catchments. Although there may be a measure of validity in this line, it seems unconvincing as an overall explanation for the abandonment of the old settlements, especially since the flow of the minor gorges which were utilized in the past is now nil in some years. Anyway, whatever the cause or the balance of factors, there has clearly been a considerable environmental change since the time when old Engaruka grew and flourished as a successful agricultural settlement during the fifteenth and sixteenth centuries; this approximate dating is established from several archaeological pointers including radiocarbon tests. From about 1700 Engaruka was deserted, since by then the effects of the changes had become sufficiently severe.

Abandonment: environmental, demographic and cultural factors

The question that remains is whether – or rather to what extent – this hydrological decline on the Rift escarpment reflects an actual reduction in rainfall in the region, as opposed to more local factors resulting from the intense land use and other activities of a concentrated population of several thousand people over those three or more centuries. The obvious consideration is whether the wood requirements – for cooking fuel, for house building, for fencing and defending the villages, and also for scaffolding and hollowed trunks to carry water along lengths of raised canals – would over this time have led to substantial deforestation of the escarpment, which in turn would have affected the aquifers and the flows of the rivers. This human factor is not easily assessed. While it may not by itself have caused a marked diminution of run-off from the highlands and the escarpment, it could have adversely affected the constancy of the discharges in the watercourses, leading to longer periods of low or nil flow broken by occasional sudden spates. A trend of this sort may indeed have occurred; at the least, as mentioned, some of the seasonal streams, including the Olemelepo, are much too temperamental to be turned to effective irrigation use now. All considered, however, it seems that in the past, at the time when old Engaruka grew and reached its most successful stage, these streams must have carried more flow in absolute annual volume, implying heavier rainfall on and above the escarpment than now. The difference may not have been so noticeable in the highlands themselves, but merely enough to maintain the flows of the lesser rivers descending the escarpment for longer seasons. Presumably, moreover, higher rainfall in the mountains towering above Engaruka implies that the Rift floor would correspondingly have been receiving somewhat more rain than now in an average year, and that in itself would have made a definite difference to cultivators there. In fact it may be that the first settlers by the Engaruka River, in the fourteenth or fifteenth centuries, managed with relatively unelaborate irrigation devices, but that the combined effects of population growth and of a slightly declining rainfall obliged the farmers to gradually develop the irrigation techniques and to increase the extent of fields relying on these. Only after some time of developing and expanding in this way would they have found themselves in the situation of having to exploit to the limit both the water resources and every hectare of land that could possibly be reached by the artificially channelled water.

Such a reconstruction of the development of the Engaruka settlement implies that, for all the special features of the system, which combined soil-conservation and fertility measures (in particular stall-feeding of livestock and manuring) with the elaborate engineering and maintenance of the irrigation of some 2,000ha of fields, in the long run it reached the limits of its technological expertise and ingenuity. From that point it would have had to contend with a decline in both the quality of the soil and the volume of the water available for irrigation. The only solution in that situation would have been for a proportion of the population to move off, whether as groups or as individuals, to settle or marry among other communities in the region. It is not clear how long this period of decline and break-up lasted. Although one might imagine that as having been very gradual, the state of the ruins – of the fields, irrigation works and villages alike, as well as of the stone enclosures and revetted clearance cairns – gives the impression of sudden and comprehensive abandonment.

That impression is not easily tested and may be mistaken. Nevertheless, such wholesale abandonment of works of this extent naturally provokes speculation about the causes of desertion. Apart from the environmental observations which have been noted, principally the diminution of the water supplies essential for the irrigation, one cannot rule out other possible contributory factors, or even catastrophic ones, such as a decimating epidemic, or an extreme El Nino event with floods and rockfalls destroying the fields and irrigation works, or again a severe eruption of the nearby volcano, Oldonyo Lengai, with the emission of more than its wonted amount of ash and dust. (Late in 1917, for instance, it is recorded that the crops of the new Engaruka settlement and the grass in the surrounding plain were devastated by a heavy fall of 'snow-like' sulphurous ash.) But on present evidence there are no definite pointers for any such catastrophic conjectures to explain the desertion which occurred about 300 years ago.

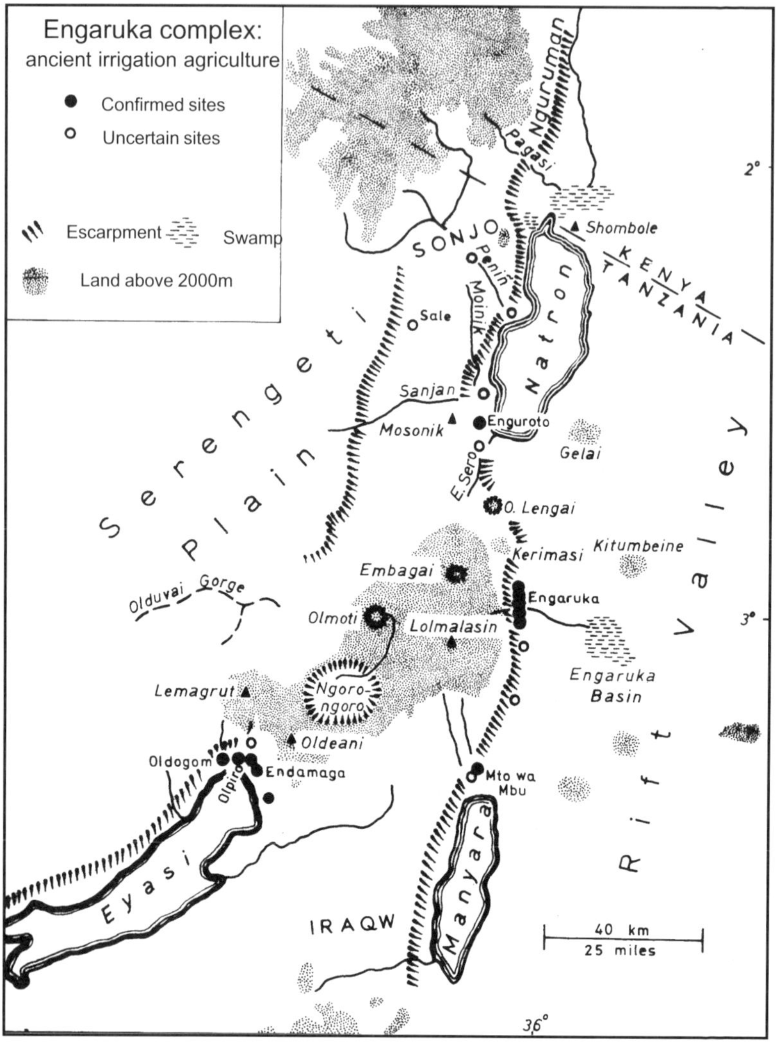

Figure 6.8 *Map: Engaruka, related Rift Valley sites and Sonjo (J.E.G. Sutton and G.M. Oteyo)*

There are lesser deserted sites in the region which are relevant to this discussion. These have remains of stone-lined fields and irrigation works which are similar, if not identical, to those of Engaruka, although in no case so extensive. These occur along the same escarpment, by Lake Natron and also towards Lake Manyara, as well as around the northeastern end of Lake Eyasi on the far side of the Crater Highlands (Figure 6.8). All these sites are in similarly low and dry situations blessed with streams flowing off the mountains and the escarpment. The similarities in both the situation and the archaeological remains, including the details of the field arrangements and their stonework, are so readily recognizable that there can be no doubt that they belonged to the same community as that of Engaruka or to closely related divisions. It is not known whether they were abandoned at exactly the same time as Engaruka or expired somewhat earlier, or whether some of them lingered on a little later. Nevertheless, this general phenomenon of desertion of specialized irrigation agricultural settlements around the Crater Highlands argues against a catastrophic local event and suggests that the main cause was an underlying environmental trend which undermined the livelihood of this series of communities. There may of course have been inherent social and cultural factors inhibiting radical adaptation to a changing situation and thus contributing to the eventual abandonment, but it is not possible to speculate on these without more specific leads.

There are few clues to where that population went, or, if one is imagining the worst scenario, where the survivors settled. This is not really a historical question, since the Engaruka system, in both a physical and a social sense, had obviously become too specially developed and integrated within its own situation to enable it to migrate as a unit and to attempt replanting itself elsewhere. Rather, the abandonment of Engaruka entailed the end of that community and identity. Elements deriving from it, however, may have become assimilated in many places. One of these was probably Sonjo, 100km north of Engaruka on the escarpment west of Lake Natron (Figure 6.8). This has consisted of several nucleated villages specializing in irrigated agriculture in river valleys and basins fed from springs, with sorghum as the main crop and *ngwara* beans an important complement. While the irrigation is an essential and integral part of the Sonjo system, dependence on it is less extreme than it was at Engaruka, as the rainfall is somewhat higher. Moreover, the fields are not divided and revetted with stone, nor are the canals and furrows stone-lined in the manner of Engaruka (the alluvial basins of Sonjo being relatively free of stone). Another difference is (or was till recently) the absence of cattle in Sonjo, for fear, so it was said, of provoking surrounding Maasai pastoralists to raid. At Engaruka cattle were eaten, though not necessarily in large numbers, as excavations in the villages have revealed; they may have been valued more for their milk and also, as argued, for their manure in the fields. The Sonjo farming system is therefore less intensively integrated, and does not necessarily derive directly from Engaruka.

But a cultural relationship is clear in the form of the nucleated villages and in certain details of the homestead platforms and their surrounds. There are, moreover, deserted village sites in Sonjo, some of them abandoned apparently one, two or more centuries ago. Fuller investigation of these deserted sites would be valuable for documenting changes in Sonjo and for establishing a sequence there, an exercise which ought to help bridge the small gap, both geographical and chronological, with Engaruka. Equally valuably it should contribute to the broader questions of environmental change in the region, as well as to the questions raised by the Engaruka evidence.

The evidence presented in this chapter, suggests a decline in rainfall in the region in about the seventeenth century, and invites us to search for other signs of such a process which might corroborate or modify that deduction. Methodologically this quest presents a difficulty since it requires much more precise dating than is currently available either for the rise, flourishing and decline of Engaruka, or for other indicators of climatic change in the ecological or archaeological record. For instance, the pottery and other artefacts obtained during excavations in the Engaruka villages are demonstrably of the later Iron Age, but they cannot be assigned to a particular century or centuries, and the several radiocarbon tests undertaken, though in broad agreement with this dating of around mid- to late second millennium AD, are by their nature insufficiently precise for charting the detailed chronology century by century.

Further north along the Rift Valley, in the elevated stretch close to the equator, there are springs which have dried out at some time in the later Iron Age. These may or may not synchronize with the Engaruka chronology. Some of these springs are by forest edges, and their expiry could have been due to deforestation or other factors in their immediate catchment rather than a decline in rainfall. In the opposite direction, in the so-called Masai Steppe lying southeast of Lake Manyara, there are sites in marginal terrain which were occupied by cultivators late in the Iron Age and subsequently abandoned. But here again the dating, as presently established, is too imprecise for pressing suggestions that these suffered from the same drier climatic trend as is suspected at Engaruka.

Clearly, these questions require a combination of approaches utilizing available techniques of interpretation and dating. One promising line is examination of lake cores in the region, with the application of OSL dating, both in the Crater Highlands above Engaruka and in the Rift. For instance, the 'Engaruka Basin', in the lowest part of this stretch of the Rift, is only some 20km distant. Normally it presents itself as bare soda flats, but in years of heavy rainfall it becomes a wide shallow lake receiving the excess discharge from the Engaruka River as well as the various seasonal and occasional streams of the district. If its sediments prove susceptible to climatic interpretation and dating, the results should relate to both Engaruka and the highlands behind.

Postscript: New Engaruka, its agriculture and changing environment

Mention was made of the existing community settled at Engaruka and its cultivation supported by irrigation. The beginnings of this new settlement can be traced to the 1890s, during the period of German rule, and its numbers were replenished on occasions during the British period, notably in the 1920s and 1940s. These settlers came to farm here from various places in north and central Tanzania, and formed (together with a few Somali traders, and a small Lutheran mission and school at the upper end) what was sometimes called the Swahili village of Engaruka; it maintained itself as an agricultural island in the midst of Maasai territory (with the escarpment to the west and Crater Highlands above gazetted within the Ngorongoro conservation area). There is no suggestion of any link with the population which had cultivated here in earlier times; the old villages on the escarpment and the associated field and irrigation system had been deserted for some two centuries previously, and during the nineteenth century only pastoral Maasai are recorded in the location.

The new settlement and its fields are on the less broken and less stony land lower down the river than the main and better preserved areas of the ancient fields (although there are clear archaeological signs that the latter did include much of the presently cultivated area). The cultivation now extends, as noted, up the Makuyuni valley and also into the lower part of the Olemelepo, the latter being fed (since about 1950) by an inter-valley canal led off the Engaruka just below its gorge and carried through the foothills (following, very roughly, the line of an ancient canal).

The composition of this community was considerably altered by the Tanzanian *ujamaa* movement of the 1970s, which involved the incorporation of Kisongo Maasai living in the surrounding area. This was more than notional villagization for administrative purposes. Fairly quickly a number of Maasai involved themselves directly in the village's politics and came to dominate it; and most are settled, if not within the nuclear part of the old village, at least close to it and to the river or the canals led off it, with the livelihood of an increasing proportion dependent on cultivation and irrigated water as much as on cattle. Officially Engaruka has been gazetted since the 1970s as two villages (*vijiji*), Engaruka Juu (upper), whose core is essentially the earlier Swahili village, and Engaruka Chini (lower), lying further into the plain. With the outlying settlements, the population runs to a few thousand.

Some of the non-Maasai moved in the 1970s to the new settlement of Selela, some 20km south along the escarpment foot, or beyond that to the expansive irrigation basin of Mto wa Mbu, on the main road linking the Mbulu Highlands (as well as Ngorongoro and Serengeti) with Arusha.

All these settlements rely for the successful continuity of their cultivation on irrigation, by means of furrows taken off streams at the foot of the escarpment, and can therefore be considered examples of intensive agriculture in the broad sense, each occupying a discrete basin on the rather arid Rift floor. But in none of them does the furrow construction compare in engineering intricacy and achievement with that of ancient Engaruka; the take-offs are lower, stone embanking is absent or minimal, and the furrows are held less close to the horizontal ideal, while the fields themselves and feeder furrows lack the meticulous stone divisions and levelling which characterize the archaeological examples. These contrasts indicate a cultural hiatus; clearly, if there has been an element of continuity in the region, it would have been indirect and in the broad principles of irrigation only, and did not occur at Engaruka itself. As argued above, it is at Sonjo in particular that cultural continuity from the old Engaruka experience seems discernible, but even there without the virtuosity of the archaeological remains of Engaruka's fields and canals. The closest replica for the artery canals constructed along the escarpment face is much further to the north in Marakwet (as described in Chapter 2).

The points of comparison and contrast here made between the irrigation works and field system of old Engaruka and those of newer settlements in the same district, including that at Engaruka itself, may cast the latter in an inferior light. Certainly, the current engineering skills and field arrangements seem less accomplished and integrated than those achieved between three and five centuries ago, and this is not entirely a case of the modern observer being mesmerized by the conspicuous archaeological record. Another feature of the old system, albeit deduced rather than directly demonstrated, would have been manuring from stall-fed cattle, a practice rarely encountered along the Rift escarpment nowadays. Nevertheless, a satisfying comparative exercise needs not only to consider the input of skill, ingenuity and labour but also to determine the output of the fields, season by season and year by year. For this purpose, the necessary observations and calculations can be taxing enough, even in the case of existing agricultural communities. For archaeological situations, with or without special features such as irrigation and manuring, sensible calculations have to contend with an array of uncertain and variable factors, ranging from the land-use and fallowing regime to the crops grown, their varieties, combinations and rotations, as well as labour availability and utilization. Ethnographic analogies can help here, but only up to a point: Sonjo and Marakwet are relevant for grain production dependent on, or supplemented by, irrigation in somewhat similar, albeit less extreme, situations. Obviously this challenge has to be taken up, for a better understanding of old Engaruka's successful development and its ultimate desertion. For the present, the impression is that the intensive labour and application of engineering perfection may not have resulted in bumper yields, but may reflect rather the increasing stress of producing an adequate harvest for a population which had grown to the maximum that could be fed off the area of irrigable land and the volume of water descending the escarpment. In this veritable island of intensive and highly specialized agriculture, the increased inputs of technology and labour may, arguably, have been rewarded eventually by decreasing output.

That is not the case with the existing settlements in the district, notwithstanding variable harvests and hardships experienced in certain years of low rainfall, when pressure on irrigation water from the limited river flows becomes more severe. Another factor now, but one which would have barely applied at old Engaruka, is the opportunity to produce crops for the regional market. This has been conspicuously developed at Mto wa Mbu in recent decades, with maize, root crops, legumes and other vegetables and fruits being produced for local and more distant sale. That opportunity is more difficult for the farmers at Engaruka, being 60km from the main road. Here maize dominates, mainly for local consumption.

Since this study has been concerned with environmental trends at Engaruka over several centuries, it is worth noting more recent changes in the vegetation alongside this brief account of the existing agricultural community. The edges of the present villages, in particular areas of fields which have been recently cultivated with help from irrigation and then abandoned, have been experiencing a marked growth of thorny weeds and bush. Moreover, in the upper parts of the ancient field area, that untouched by recent settlement and cultivation, the thornbush cover has become noticeably thicker between the 1960s and the 1990s. This is illustrated by photographs taken of various parts of the ancient field remains and of particular features, notably the embanked causeway of the great northern canal, at intervals since the late 1950s (Figure 6.6 dates to 1972), and also by the air-photographic mosaic of 1964. That, if repeated now, would show very much more bush covering considerable areas, and would be much less revealing about the field and furrow arrangements and the archaeological features generally. Likewise, the photograph of 1971 of the irrigation grid below the intermediate north gorge (Figure 6.7) cannot be repeated now. The stone lines and divisions are still there but are masked by the bush.

Although the growth of bush and sturdy weeds around the settlements and on recently abandoned cultivation plots can be attributed to the soil disturbance caused by both hoeing and by concentrations of people and goats, as well as a few cattle and donkeys kept there, the cause of the change in the uninhabited upper ruin area is less clear. Goats browse here and a few cattle are brought to graze especially after rain, but this area has not suffered obvious over-grazing. The main factor may be the opposite, that of reduced exploitation of the grass in this particular area by Maasai herds (because of restricted access owing to the numbers of settled cultivators nearby) and therefore of less burning after grazing. At the same time there has been a reduction in the numbers of gazelles, impala, zebra and giraffe. They still have access to the river, the escarpment base and the foothills, but, because of a bigger human population than previously along the Engaruka River and in the Makuyuni valley, that access is not so unfettered. Moreover, with a reduction of grass and more bush, some of the herbivores, the gazelles and zebra in particular, are less attracted. And of course rhinos – which it is claimed would have been especially effective in holding back bush – have not been seen below the escarpment for some time.

These observations may or may not be pertinent to the explanation of this vegetational trend in the second half of the twentieth century. Moreover, the change itself has not been monitored methodically, but is merely recorded impressionistically from memories and from dated photographs taken for other purposes. It is an environmental change – or more probably part of a broader set of developments in this location – which deserves more systematic ecological investigation. That will doubtless assist some of the insights needed for a more sophisticated evaluation of the trends and changes in the history of Engaruka and its agriculture over half a millennium, which is the main subject of this chapter.

Bibliographic Notes

The most recent account of Engaruka, including the survey of the fields and irrigation system, the results of excavations (of 1935, 1964-6 and 1982) and the dating evidence, is my site study, 'Engaruka: an irrigation agricultural community in northern Tanzania before the Maasai' (*Azania*, 33 1998: 1-37).

More particularly on the fields, irrigation systems and agricultural techniques: 'Engaruka and its waters' (*Azania* 13, 1978: 37-70); 'Irrigation and soil-conservation in African agricultural history' (*Journal of African History* 25, 1984: 25-41); 'The irrigation and manuring of the Engaruka field system' (*Azania* 21, 1986: 27-51); and 'Towards a history of cultivating the fields' (*Azania* 24, 1989: 99-112).

On Sonjo see W.M. Adams, T. Potkanski and J.E.G. Sutton, 'Indigenous farmer-managed irrigation in Sonjo' (*Geographical Journal* 160, 1994: 17-32); also D. Nurse and F. Rottland, 'The history of Sonjo and Engaruka: a linguists' view' (*Azania* 28, 1993: 1-5).

The agriculture of the 'new' Engaruka villages, and also that of Selela and Mto wa Mbu to the south, is described in a report by Pernille Bertelsen, *Farmer-managed irrigation systems, Manyara Division* (Institute of Development and Planning, Aalborg University, 1995).

Seven

When Islands Expand
Intensification & Sustainability

WILLIAM M. ADAMS

Introduction

This book has described a series of agricultural production systems in Tanzania, Kenya and Ethiopia as islands, because of the way in which they have been set apart from the systems that surround them. Agriculture in these islands is distinctive, intensive and specialized. Production systems in the seas that surround them have often involved a far greater dependence on livestock, and a much more partial (and less labour-demanding) dependence on agriculture, although it may also be specialized, in that it also involves highly developed technological complexity, ecological soundness and flexibility (Sutton 1989). Flexibility and diversity have been highlighted repeatedly as critical elements of livelihood strategies in sub-Saharan Africa, particularly in drier areas and among livestock keepers (Anderson and Johnson 1988, Scoones 1995).

It is, however, the intensiveness of the agriculture in these islands that has attracted the attention of scholars in the past, and which is the basis of their fascination at the present time. The simplest definition of agricultural intensification is a sustained increase in the amount of labour applied to each hectare of land (intensity of the use of labour); a recent elaboration in the East African context (Tiffen et al. 1994: 29) expresses it as 'increased average inputs of labour or capital on a smallholding, either on cultivated land alone, or on cultivated and grazing land, for the purpose of increasing the value of output per hectare'.

Agricultural change

The reason for the interest in intensive agricultural systems in Africa is that in the past they have been seen as so unusual. In the time-frame of many centuries, sub-Saharan Africa has been land-rich and labour-scarce (Iliffe 1995). Faced with the huge task of stopping the regeneration of wild plants on farmland and maintaining soil fertility, most African farmers have chosen to move, clear and burn bush, and start again. Rural Africans struggled for millennia with a limited supply of labour, and mostly evolved strategies for environmental management and for subsistence that could be effective with low labour inputs.

Faced with a predominance of labour-saving extensive agricultural systems and lacking historical insight, observers of agriculture in Africa during the twentieth century reacted in two ways. The first was a Malthusian concern about an incipient sustainability crisis. The second was a productivist concern about farmer conservatism. The turnaround in demographic fortunes, and the high population growth that became so prominent in the twentieth century, persuaded many observers to make a stark Malthusian analysis of the possibility of African agricultural production keeping pace with population and to emphasize the likelihood of severe environmental degradation (Harroy 1949; see also Anderson 1984). The advent of economic planning (from the 1940s, and particularly as African countries gained independence from the late 1950s onwards) led to development strategies that looked to agriculture to yield economic surplus for investment elsewhere (in social and economic infrastructure or industry). Agricultural policy demanded increased economic output from farmers and engagement in the commercial sector (preferably in cash crops that could be exported). From that imperative grew the conventional thinking about agricultural development based on modern agricultural techniques (inorganic fertilizer, pesticides, newly bred seed varieties, mechanization) and market orientation. In certain places this approach took hold; elsewhere farmers were unimpressed with the balance of risks and opportunities offered by the modern state and economy, and resistant to their demands. In Hyden's (1980) term, the peasantry remained determinedly 'uncaptured'.

Both the Malthusian analysis of impending disaster and the developmental analysis of the urgent need for production increase presented African agriculture as resistant to change. Underlying both these attitudes there was the broader assumption that African societies had been relatively unchanging over time. Anderson and Johnson (1988: 2) comment that the study of Africa has often suffered from an assumption of timelessness, and Sutton (1990: 2) warns that in much historical writing and debate 'there persists the notion of an essentially unchanging, a supposedly "traditional" African past, the details of which need little researching, since they can be readily assumed'. The dominant view was of African agriculture being almost everywhere locked in a cycle of low-input and low-output, and African farmers being resistant to change – uninnovative, unspecialized, uninterested in economic relations and unresponsive to economic and environmental change. There were exceptions to this view (de Schlippe 1956, Allan 1965), but it remained dominant for much of the century, and still retains its hold over the most recent development planning.

To most twentieth-century observers, intensification was therefore something novel for African farmers, and something into which, moreover, they needed to be persuaded or indeed forced through development projects, technologies, incentives and fiscal pressures. Intensification was something inherent to modern states and capitalist economies, and a process that had to be driven by outsiders, against farmers' wishes but for their own good. However, as the contributors to this volume amply illustrate, these are deeply erroneous views. It is precisely the dynamism of agricultural systems over time, particularly intensive systems, that is their most remarkable feature. Farmers have a very acute sense of what return they may obtain from different kinds of techniques, and they will innovate rapidly if the new ideas and the technologies they offer actually work. It is a matter of record that, through the second half of the twentieth century, many did not. Despite decades of development planning, much agriculture remains dependent on local knowledge and capacity.

Indigenous intensification

By the end of the twentieth century new studies began to suggest that under certain circumstances, intensification could take place without external modern intervention. This overturned previous dismissive assessments of the sustainability and productive potential of African agriculture which had been dominant in the colonial period (see Richards 1985, 1986), and which were shared by Marxist scholars and agricultural modernizers in the third quarter of the twentieth century. Slowly an alternative view has gained prominence, one which sees African farmers as skilful managers of the land, knowledgeable in their decisions about what to grow and how, self-conscious in their decisions relating to sustainability, and subtle in their responses to environmental and economic opportunities and threats (Richards 1985, Netting 1993, Mortimore 1998). Malthusian models of society and environment in sub-Saharan Africa have been challenged (Roe 1991, Leach and Mearns 1996), and there has been renewed interest in the ideas of Boserup (1965) about the potential for agricultural intensification in conditions of increasing population density.

Alongside this has emerged a broad recognition of the possibility that growing densities of rural people can have positive consequences for the environment and economy, and reviews suggest that African agriculture has been showing consistent adaptability and output growth (Wiggins 1995). Studies of smallholders reveal their capacity to make decisions and to deal with difficult economic and environmental conditions. In the case of the Kano Close-Settled Zone (KCSZ) and nearby areas of Nigeria and Niger (Mortimore 1989, Mortimore and Adams 1999), and also in that of Machakos District in Kenya (Tiffen and Mortimore 1992, 1994, Tiffen et al. 1994, Mortimore and Tiffen 1995), it is being argued that, under the right conditions, agricultural intensification can take place indigenously, without external stimulation.

It is not inevitable that rising human populations are accompanied by declines in production and soil fertility. At low population densities such declines may not matter (because extension of the cultivated area remains an option), and may not be easily reversed (because there is too little labour to invest in intensive agricultural practices). However, at a certain level of population density it may become less feasible to extend cultivation and more attractive to intensify. At this point, one might expect agricultural intensification to take place, as farmers make investments in the productive potential of their land.

Small farmers can and will invest in their land as populations rise, creating 'landesque capital' (Brookfield 1984). Conventional views of capital investment by smallholder farmers have emphasized the limited capacity for formal capital investment; there are, however, various ways in which smallholder farmers create capital (Mortimore 2002). These include investments by family or other labour in fields (boundaries and terraces, irrigation structures, tree planting and protection, the manufacture of manure or compost and its application to soil, field ridging or other tillage improvements) and in homesteads (food-storage structures or livestock enclosures). They also include investments in the living productive resources of the farm (such as acquiring and managing seed varieties, tools or animal harnesses, purchase or breeding of draught, milking or meat-providing animals), and investments in the productive capacity of household members (such as in children's education).

Investments of this kind tend to be small and episodic, the piecemeal fixing of separate facies of a sustainable livelihood strategy into a larger mosaic (Mortimore 2002). Furthermore, progress is not always enduring, as the demands of a household's own life cycle (births, deaths, marriages) and external crises (drought, warfare, pest attack) draw down or wipe out the capital investments of preceding years. Sutton's account of the abandoned irrigation system of Engaruka (see Chapter 6) graphically demonstrates that intensive production systems are by no means always sustainable, although the exact cause of the eventual downfall of Engaruka remains unclear. However, it is now recognized that households do invest in this way when they can, and that indigenous intensification has been a much more common feature of African agricultural systems in the twentieth century than was previously thought.

Why intensify?

Agricultural modernizers in Africa have therefore been mistaken in seeing intensification as an unprecedented and unusual process. Rather, it should be seen as a process that is only worthwhile under certain conditions, and historically these have not been widespread. Sutton (1989) points out that it is a reasonable assumption that extensive agricultural systems took the form they did because they were more efficient in their use of available labour in relation to the returns needed from the land. He notes the effort of intensification, pointing out that more laborious farming methods are only sensibly undertaken 'if both the needs and the expectation of rewards are overwhelming' (Sutton 1989: 100). Intensification is not necessarily an obvious, easy or attractive choice for smallholder farmers.

The question discussed by the authors in this book in a historical context is exactly that which is being asked by development planners in Africa at the start of the twenty-first century: why do farmers intensify? The answer, inevitably, turns out to be complicated, and locally contingent. Four factors are crucial: the ecology of the production system; the nature of the household labour economy; the relationship between cultivation and wider cultural practices; and the nature of exogenous forcing factors.

The ecology of the production system is important because it sets certain constraints both on crop choice and cropping system diversity and on the range of non-farm productive opportunities. The greater the variety of crops, and the greater the variety of kinds of a single crop, the more opportunities there are for farmers to make subtle adjustments for soil conditions and also to respond to environmental variability by an appropriate crop choice. In studies in northeastern Nigeria a range of millet and sorghum varieties was recorded, for instance, in the KCSZ (see above) which were combined with other crops (especially cowpeas) in a wide-ranging cropping system adapted to individual farmer's choices. In drier villages new crops such as the cow melon (*guna*) were introduced into cropping portfolios following the droughts of the 1970s; these were planted as the rains ended and provided the chance of a crop in the event of failure of the main crop, and also an available source of edible oil (Mortimore and Adams 1999). Diversity and choice of this sort are integral to the possibility of intensification. The opportunity of integrating livestock and crops in a single household economic system and a closed-loop nutrient cycle (Harris 1999) is also important. Several studies in this book have emphasized the integration of livestock and crop farming in one economic system, notably the case of an Iraqw household (as described by Loiske in Chapter 5).

Second, the attractiveness of more intensive cultivation techniques will depend on the availability of family labour, and the relative economic returns from on-farm as opposed to off-farm activities. Netting (1993) stressed the economic basis of smallholder decision-making, pointing further that since in the past many households were remote from the modern economy (and even now may remain so), an economic calculus is fundamental to the allocation of labour to different tasks. If livestock-keeping offers better returns than intensifying agriculture, farmers will be episodic or permanent agro-pastoralists. Farming households will allocate different amounts of labour to different fields, and make crop choices accordingly (Mortimore and Adams 1999). As the availability of household labour changes, decisions taken about where effort is to be invested will likewise change. (For instance, the availability or non-availability of teenage boys will influence which jobs are handled and which neglected.) Different households will make contrasting choices about the intensity of cropping; thus a large household with plentiful male labour might make quite different decisions about the intensity of production from those of an adjacent female-headed household without such a resource.

Third, the attractiveness of intensification will depend on the relationship between cultivation and wider cultural practices. Farming is not merely a means of producing a livelihood and generating surplus or wealth; rather it is affected by, and is also central to, identity and social relations. This is clear from studies of Maa-speaking cattle-keepers and farmers, for whom the possibility of shifting from one economy to the other was a key element in survival strategies at various times in the last two centuries (Anderson 1988, Spear 1993, 1997). In this volume, Östberg (Chapter 2) argues convincingly that for the Marakwet the irrigation system was an element in the formation of ethnic identity, thus opposing the simplistic view that it was an intensive agricultural technology they adopted. Watson's paper (Chapter 3) demonstrates the links between social organization, the control of labour and the nature of agricultural activity in Konso: in her interpretation the terraced field systems are literally a reflection of social relations.

Finally, the move towards more intensive agriculture must reflect external factors of environment, economy and security. Interdisciplinary communication, and particularly the combination of palaeoenvironmental research with oral history, is starting to reveal something of the complex interplay of culture, environment and economy in Africa in recent centuries. Agriculture exists within an economy, and responds to opportunities within that economy. Explanations of intensification therefore have to take account of the economic context. Netting (1993), for example, describes the way in which the Kofyar abandoned intensive methods when they moved off the Jos Plateau in Nigeria to the relatively land-plentiful plains, but reverted to more intensive methods when the new frontier in turn began to close. Similarly, the intensity of cultivation in Iraqw varies with population density (Chapters 4 and 5). The studies in this book of both Marakwet and Iraqw attach considerable importance to economic networks between kin and between neighbours. Agriculture and trade are parts of a linked system of production and exchange, and intensification must both reflect and in turn contribute to that exchange and economic life. A process of intensification therefore feeds to an extent on itself as a virtuous circle: as Börjeson comments (Chapter 4), agricultural intensification can be 'its own driving force'.

Researching agricultural islands

The farming systems discussed in this book, at Engaruka and in Iraqw, Marakwet and Konso, are not artefacts of academic enquiry, but are remarkable and very clear features on the ground. To visit the irrigated fields of Marakwet adjacent to the dry bush of the floor of the Rift Valley, or those of Sonjo after the open bush of the Salei Maasai, is to experience them, as the title of this book implies, as islands of intensification. Furthermore, not only are these islands economically, culturally and ecologically distinct; equally important, as the research presented here also demonstrates, they have been remarkably persistent through time, i.e. they are enduring features of the East African landscape. As Börjeson demonstrates (in Chapter 4), the landscapes of Iraqw that Baumann encountered and drew in the 1890s are recognizable around a century later. In the Marakwet case Östberg's (Chapter 2) hypothesis is the slow co-evolution of the farming system and cultural identity.

The existence of these islands has a number of implications for our understanding of African agriculture. First, this research shows that it is necessary to abandon the notion that agriculture in sub-Saharan Africa is somehow homogeneous in space. This view, though long abandoned by scholars, retains its power among technical experts, politicians and lay commentators. It is clear from these East African case studies that farmers manage land in quite specific ways in the light of economic and environmental opportunities. Farmers do different things in different places, and African agriculture is highly specialized for particular local environmental conditions. This is demonstrated here for a series of quite different agricultural systems; studies of other regions, for instance the Sahel (Raynaut et al. 1998, Mortimore and Adams 1999), bear out the same point.

A second implication of this work is that in order to understand agriculture in Africa we need to engage in detailed locally specific research and case studies. If agricultural systems are the product of decisions responding to local cultural, social, economic and environmental conditions, then research into those systems must engage with all these issues. In other words, research into agriculture must involve itself in far more than agronomy or economics; it has to be multidisciplinary, in particular including anthropology and history. The disciplinary boundaries between these modes of academic enquiry have been erected in response to the political economy of universities in the twentieth century, but they do not make much sense in the context of African farming households. Researchers, like farmers, need to be broad-minded, multiskilled and open to the opportunities and threats posed by environmental and institutional change. The case studies reported here by Börjeson, Loiske, Östberg and Watson provide models for the conduct of intensive local observation and research.

The third implication of the work discussed in this book is that in order to understand agriculture and agricultural change, research must take a broad framework in both space and time. In space, there is a need to understand agriculture and intensification at the landscape scale, seeking and analysing the connections between different locations, between core and periphery, through kinship and economic exchange, as Börjeson and Loiske do for Iraqw. We need to understand the temporal evolution of cultural practices and economic relations, as both the political economy and the environment change, whether at the intergenerational scale (as in Iraqw), or over longer time-scales (as Östberg describes for Marakwet). Thus, the influential study of agricultural change by Tiffen, Mortimore and their Kenyan colleagues in Machakos was effective precisely because they took a historical and geographical approach to the area, and drew on a wide range of data sources from which to construct a coherent story. Certain other recent East African studies, ranging in space from the same district to Kigezi in Uganda (Rocheleau et al. 1995, Murton 1999, Lindblade et al. 1998), have followed to advantage a similar historically and geographically literate approach.

There are urgent research priorities here. Attempts to identify a route towards sustainable agriculture in contemporary Africa have much to learn from the past, and specifically from studies of the kind discussed in this volume. Such learning is vital to avoid development interventions that reduce the capacities of rural smallholders to cope with the multitude of environmental and political and economic challenges that face them and hinder their efforts to achieve sustainable livelihoods.

References

Adams, W.M., 1989. 'Definition and development in African indigenous irrigation', *Azania* 24: 21-7.

—— 1996. 'Irrigation, erosion and famine. Visions of environmental change in Marakwet, Kenya', in M. Leach and R. Mearns (eds) *The Lie of the Land: Challenging Received Wisdom on the African Environment*, 155-67. Oxford: James Currey for the IAI.

Adams, W.M. and Anderson, D.M. 1988. 'Irrigation before development', *African Affairs* 87: 519-35.

Adams, W.M., Potkanski, T. and Sutton, J.E.G. 1994. 'Indigenous farmer-managed irrigation in Sonjo, Tanzania', *Geographical Journal* 160: 17-32.

Adams, W.M., Watson E.E. and Mutiso, S.K. 1997. 'Water, rules and gender: water rights in an indigenous irrigation system, Marakwet, Kenya', *Development and Change* 28: 707-30.

Allan, W. 1965. *The African Husbandman*. Edinburgh: Oliver and Boyd.

Ambler, C.H. 1988. *Kenyan Communities in the Age of Imperialism: the Central Region in the Late Nineteenth Century*. New Haven, CT: Yale University Press.

Amborn, H. 1989. 'Agricultural intensification in the Burji-Konso cluster', *Azania* 24: 71-84.

Ambrose, S.H. 1982. 'Archaeology and linguistic reconstructions of history in East Africa', in C. Ehret and M. Posnansky (eds) *The Archaeological and Linguistic Reconstruction of African History*, 104-57. Berkeley: University of California Press.

Anderson, D.M. 1984. 'Depression, dust bowl, demography and drought: the colonial state and soil conservation in East Africa during the 1930s', *African Affairs* 83: 321-44.

—— 1988. 'Cultivating pastoralists: ecology and economy among the Il Chamus of Baringo, 1840–1980', in D.H. Johnson and D.M. Anderson (eds) *The Ecology of Survival: Case Studies from Northeast African History*, 241-60. London and Boulder, CO: Crook.

—— 1989. 'Agriculture and Irrigation Technology at Lake Baringo in the Nineteenth Century', *Azania* 24: 84-97.

Anderson, D.M. and Johnson, D.H. 1988. 'Ecology and Society in Northeast African History', in D.H. Johnson and D.M. Anderson (eds) *The Ecology of Survival: Case Studies from Northeast African History*, 1-24. London and Boulder, CO: Crook.

Appadurai, A. 1986. 'Commodities and the politics of value', in A. Appadurai (ed.) *The Social Life of Things: Commodities in Cultural Perspective*, 3-63. New York: Cambridge University Press.

Arens, W. 1979. *On the Frontier of Change: Mto wa Mbu, Tanzania*. Ann Arbor: University of Michigan Press.

Austin, H.H. 1973 (1903). *With Macdonald in Uganda* (new introduction by A.T. Matson). Folkestone and London: Dawsons.

Bagshawe, F.J. (c. 1920). 'Tribal studies'. Unpublished manuscript (Mss.Afr.s.301),

Rhodes House Library, Oxford.
—— 1926. 'Peoples of the Happy Valley, East Africa', Part IV, *Journal of the Royal African Society* 25: 59-74.

Bart, F., Charlery de la Masselière, B. and Calas, B. 1998. *Caféicultures d'Afrique orientale: territoires, enjeux, politiques.* Paris: Editions Karthala.

Barth, F. 1969. 'Introduction', in F. Barth (ed.) *Ethnic Groups and Boundaries: the Social Organization of Cultural Difference*, 9-38. Bergen and Oslo: Universitetsforlaget; London: Allen and Unwin.

Baumann, O. 1894. *Durch Massailand zur Nilquelle: Reisen und Forschungen der Massai-Expedition des deutschen Antisklaverei-Komite in den Jahren 1891–1893.* Berlin: Reimer.

Bebbington, A. 1997. 'Social capital and rural intensification: local organisations and islands of sustainability in the rural Andes', *Geographical Journal* 163: 189-97.

Beech, M.W.H. 1911. *The Suk, their Language and Folklore.* Oxford: Clarendon.

Behnke, Jr R.H., Scoones, I. and Kerven, C. (eds) 1993. *Range Ecology at Disequilibrium. New Models of Natural Variability and Pastoral Adaptation in African Savannas.* London: ODI.

Beidelman, T.O. 1960. 'The Baraguyu', *Tanganyika Notes and Records* 55: 245-78.

Berger, P. and Kiessling, R. 1998. *Iraqw Texts* (collected by Paul Berger, translated, revised and edited by Roland Kiessling). Archiv Afrikanistischer Manuskripte, Cologne: Köppe.

Berntsen, J.L. 1976. 'The Maasai and their neighbours: variables of interaction', *African Economic History* 2: 1-11.

Berry, S. 1993. *No Condition is Permanent: the Social Dynamics of Agrarian Change in Sub-Saharan Africa.* Madison: University of Wisconsin Press.

Bertelsen, Pernille. 1995. *Farmer-managed Irrigation Systems, Manyara Division.* Institute of Development and Planning, Aalborg University.

Blaikie, P. and Brookfield, H. 1987. *Land Degradation and Society.* London: Methuen.

Blaut, J. 1993. *The Colonizer's Model of the World: Geographical Diffusionism and Eurocentric History.* New York: Guilford Press.

Bollig, M. 1990. 'An outline of pre-colonial Pokot history', *Afrikanistische Arbeitspapier* 23: 73-91.

—— 1993. 'Intra- and interethnic conflict in Northwest Kenya: a multicausual analysis of conflict behaviour', *Anthropos* 88: 176-84.

—— 1998. 'Moral economy and self-interest: kinship, friendship, and exchange among the Pokot (N.W. Kenya)', in T. Schweizer and D.R. White (eds) *Kinship, Networks, and Exchange*, 137-57. Cambridge University Press.

—— 2000. 'Staging social structures: ritual and social organisation in an egalitarian society. The pastoral Pokot of northern Kenya', *Ethnos: Journal of Anthropology* 65: 341-65.

Börjeson, L. 1998. *Landscape, land use and land tenure in Mama Issara, Tanzania.* Minor Field Studies No. 47. Uppsala: Swedish University of Agricultural Sciences.

—— 1999. 'Listening to the land: the Iraqw intensive farming system as told by a hill and its inhabitants', in M. Widgren and J.E.G. Sutton (convenors) *'Islands' of Intensive Agriculture in the East African Rift and Highlands: a 500-year Perspective*, 56-73. Working paper No. 43, EDSU, Stockholm University.

Boserup, E. 1965. *The Conditions of Agricultural Growth.* London: Allen & Unwin.

Braun, R. 1990. *Environmental Impact Assessment of Mto wa Mbu, Tanzania.* Geneva:

ILO, Irrigation and Flood Control Project.

Bronson, B. 1972. 'Farm labour and the evolution of food production', in B. Spooner (ed.) *Population Growth: Anthropological Implications,* 190-218. Cambridge, MA: Massachusetts Institute of Technology.

Brookfield, H.C. 1972. 'Intensification and disintensification in Pacific agriculture: a theoretical approach', *Pacific Viewpoint* 13: 30-48.

—— 1984. 'Intensification revisited', *Pacific Viewpoint* 25: 15-44.

—— 1986. 'Intensification intensified', *Archaeology in Oceania* 31: 177-80.

Brumfiel, E.M. and Earle, T.K. 1987. *Specialization, Exchange and Complex Societies.* Cambridge University Press.

Cappon, J.C. 1985. *District Atlas, Elgeyo Marakwet.* Iten (Kenya): Ministry of Planning and National Development.

Chaundy, G.H. 1939. 'Primitive agricultural methods of the West Suk tribe and some improvements', *East African Agricultural Journal* 5: 23-30.

Conelly, W.T. 1994. 'Population pressure, labor availability, and agricultural disintensification: the decline of farming on Rusinga Island, Kenya', *Human Ecology* 22: 145-70.

Conelly, W.T. and Chaiken, M.S. 2000. 'Intensive farming, agro-diversity, and food security under conditions of extreme population pressure in western Kenya', *Human Ecology* 28: 19-51.

Conte, C. 1994. 'Transformations along the gradient: ecological change in the mountains and plains of northeastern Tanzania's West Usambara mountains, c. 1860–1970', PhD thesis, Department of History, Michigan State University.

—— 1996. 'Nature reorganized. Ecological history in the plateau forests of the west Usambara mountains 1850–1935', in Maddox et al. (eds) *Custodians of the Land: Ecology and Culture in the History of Tanzania*, 96-121. London: James Currey.

Critchley, W. 1979. 'Chesongoch Agricultural Project 1977–1979, final report'. Mimeo: Chesongoch Catholic Mission.

Dahl, G. and Hjort, A. 1976. *Having Herds.* Stockholm Studies in Social Anthropology, No. 2, Stockholm University.

Dahlberg, A.C. 1996. *Interpretations of environmental change and diversity: a study from North East District, Botswana.* Dissertation Series, No. 7, Department of Physical Geography, Stockholm University.

De Langhe, E., Swennen, R. and Vuylsteke, D. 1995. 'Plantain in the early Bantu world', in J.E.G. Sutton (ed.) *The growth of farming communities in Africa from the equator southwards. Azania* 29-30: 147-60.

Dietz, T. 1987. *Pastoralists in Dire Straits. Survival Strategies and External Interventions in a Semi-arid Region at the Kenya/Uganda Border: Western Pokot, 1900–1986.* Amsterdam: University of Amsterdam.

Donham, D. 1985. *Work and Power in Maale, Ethiopia*. Michigan: UMI Research Press.

—— 1990. *History, Power, Ideology: Central Issues in Marxism and Anthropology*. Cambridge University Press.

Doolittle, W.E. 1984. 'Agricultural change as an incremental process', *Annals of the Association of American Geographers* 74: 124-37.

Edgerton, R.B. 1971. *The Individual in Cultural Adaptation: a Study of four East African Peoples.* Berkeley: University of California Press.

Ehret, C. 1998. *An African Classical Age: Eastern and Southern Africa in World History, 1000 B.C. to 400 A.D.* Charlottesville: University Press of Virginia.

—— 2002. *The Civilizations of Africa: a History to 1800.* Oxford: James Currey.

Eliot, C. 1911. 'Introduction', in M.W.H. Beech, *The Suk, their Language and Folklore,* xi-xxiv. Oxford: Clarendon.

Fairhead, J. and Leach, M. 1996. *Misreading the African Landscape: Society and Ecology in a Forest-savanna Mosaic.* Cambridge University Press.

Feierman, S. 1990. *Peasant Intellectuals: Anthropology and History in Tanzania.* Madison: University of Wisconsin Press.

Fleuret, P. 1985. 'The social organisation of water control in the Taita Hills, Kenya', *American Ethnologist* 12: 103-18.

Fosbrooke, H.A. 1954. 'The defensive measures of certain tribes in north-eastern Tanganyika. Part II: Iraqw housing as affected by inter-tribal raiding', *Tanganyika Notes and Records* 36: 50-7.

—— 1955. 'The Mbulu Highlands: problems of people, land and cattle'. Unpublished manuscript, University Library, University of Dar es Salaam.

Friedman, J. 1975. 'Tribes, states and transformations', in M. Bloch (ed.) *Marxist Analyses and Social Anthropology*. London: Malaby Press.

George, K.M. 1999. 'Objects on the loose: ethnographic encounters with unruly artefacts. A foreword', *Ethnos* 64: 149-50.

Gillman, C. 1936. 'A population map of Tanganyika territory', *Geographical Review* 26: 353-73.

Gourou, P. 1948. *Les pays tropicaux.* Paris: Presses universitaires de France.

—— 1991. *L'Afrique tropical, nain ou géant agricole?* Paris: Flammarion.

Gray, R.F. 1963. *The Sonjo of Tanganyika: an Anthropological Study of an Irrigation-based Society.* London: Oxford University Press for IAI.

Grove, A. 1993. 'Water use by the Chagga on Kilimanjaro', *African Affairs* 92: 431-48.

Grove, A.T. and Sutton, J.E.G. 1989. 'Agricultural terracing south of the Sahara', *Azania* 24: 98-112.

Gunnel, Y. 1997. 'Comparative regional geography in India and West Africa: soils, landforms and economic theory in agricultural development strategies', *Geographical Journal* 163: 38-46.

Gupta, A. and Ferguson, J. 1992. 'Beyond "culture": space, identity, and the politics of difference', *Cultural Anthropology* 7: 6-23.

Hakansson, N.T. 1989. 'Social and political aspects of intensive agriculture in East Africa', *Azania* 24: 12-20.

—— 1994. 'Grain, cattle and power: social processes of intensive cultivation and exchange in precolonial Western Kenya', *Journal of Anthropological Research* 50: 249-76.

—— 1995. 'Irrigation, population pressure, and exchange in precolonial Pare, Tanzania', *Research in Economic Anthropology* 16: 297-323.

—— 1998. 'Rulers and rainmakers in precolonial South Pare, Tanzania: exchange and ritual experts in political centralization', *Ethnology* 37: 263-84.

Hallpike, C.R. 1968. 'The status of craftsmen among the Konso of south-west Ethiopia', *Africa* 28: 3.

—— 1972. *The Konso of Ethiopia: A Study of the Values of a Cushitic People.* Oxford University Press.

Harris, F.M.A. 1999. 'Nutrient management strategies of smallholder farmers in a short-fallow farming system in north-east Nigeria', *Geographical Journal* 165: 275-85.

Harroy, J.-P. 1949. *Afrique: terre qui meurt: la dégradation des sols Africains sous l'influence de la colonisation*. Brussels: Marcel Hayez.

Hartley, B.J. 1938. 'An indigenous system of soil protection', *East African Agricultural Journal* 4: 63-6.

Hecklau, H. 1970. 'Bewässerungsfeldbau in Kenya', in K.A. Boesler and A. Kuhn (eds) *Aktuelle Probleme geographischer Forschung*, 475-92. Abhandlungen des 1. Geographischen Instituts der freien Universität Berlin.

Hennings, R.O. 1951. *African Morning*. London: Chatto and Windus.

Hodder, I. 1982. *Symbols in Action: Ethnoarchaeological Studies of Material Culture*. Cambridge University Press.

Höhnel, L.R. von, 1892. *Zum Rudolph-See und Stephanie-See. Die Forschungsreise des Grafen Samuel Teleki in Ost-Aequatorial-Afrika 1887–1888*. Vienna: Hölder.

Holtland, G. 1994. 'A farming system analysis of Mvumi Division, Dodoma Region, Tanzania'. Tanzania: Mvumi Rural Training Centre communication bulletin No. 1.

Hyden, G. 1980. *Beyond Ujamaa in Tanzania: Underdevelopment and an Uncaptured Peasantry*. London: Heinemann.

Igbokwe, E.M. 1996. 'A SWC system under threat: a visit to Maku, Nigeria', in C. Reij, I. Scoones and C. Toulmin (eds) *Sustaining the Soil: Traditional Soil and Water Conservation in Africa*, 219-27. London: Earthscan.

Iliffe, J. 1979. *A Modern History of Tanganyika*. Cambridge University Press.

—— 1995. *Africans: the History of a Continent*. Cambridge University Press.

Jaeger, F. 1913. *Das Hochland der Riesenkrater und die umliegenden Hochländer Deutsch Ostafrikas, Teil II: Länderkundliche Beschreibung*. Berlin: Mitteilungen aus den deutschen Schutzgebieten, Ergänzungsheft, Nr. 8.

Jensen, A.E. 1960. 'Feld-Terrassen und Megalithen', *Paideuma* 7: 258-73.

Kagwanja, M. 1998. *Killing the Vote. State Sponsored Violence and Flawed Elections in Kenya*. Nairobi: Kenyan Human Rights Commission.

Kahn, J.S. 1981. 'Marxist anthropology and segmentary societies: a review of the literature', in J.S. Kahn and J.R. Llobera (eds) *The Anthropology of Pre-capitalist Societies*, 57-88. London: Macmillan.

Kassogué, A., Momota, M., Sagara, J. and Schutgens, F. 1996. 'A measure for every site: traditional SWC techniques on the the Dogon Plateau, Mali', in C. Reij, I. Scoones and C. Toulmin (eds) *Sustaining the Soil: Traditional Soil and Water Conservation in Africa*, 69-79. London: Earthscan.

Kenya, Republic of, National Assembly. 1992. *Report of the Parliamentary Select Committee to Investigate Ethnic Clashes in Western and Other Parts of Kenya 1992*.

Kimambo, I.N. 1996. 'Environmental control & hunger in the mountains & plains of nineteenth-century northeastern Tanzania', in G. Maddox, J.L. Giblin and I.N. Kimambo (eds) *Custodians of the Land. Ecology and Culture in the History of Tanzania*, 71-95. London: James Currey.

Kipkorir, B. (with F.B. Welbourn). 1973. *The Marakwet of Kenya: a Preliminary Study*. Nairobi: East African Literature Bureau.

Kjekshus, H. 1996 (1st edn 1977). *Ecology Control and Economic Development in East African History*, 2nd edn. London: James Currey.

Koponen, J. 1988a. *People and Production in Late Precolonial Tanzania*. Helsinki: Finnish Society for Development Studies.

—— 1988b. 'War, famine and pestilence in late precolonial Tanzania: a case for a heightened mortality', *International Journal of African Historical Studies* 21: 637-76.

Kopytoff, I. 1986. 'The cultural biography of things: commoditization as process', in A. Appadurai (ed.) *The Social Life of Things: Commodities in Cultural Perspective*, 64-91. New York: Cambridge University Press.

Kurita, K. 1983. 'Market at Chesegon village: a preliminary report on economic activities of the Pokot and the Marakwet', in B.E. Kipkorir, R.C. Soper and J.W. Ssennyonga (eds) *Kerio Valley: Past, Present and Future*, 61-74. University of Nairobi, Institute of African Studies.

Lamphear, J. 1988. 'The people of the grey bull: the origin and expansion of the Turkana', *Journal of African History* 29: 27-40.

Lawi, Y.Q. 1992. '"Modernization" and the de-harmonization of the man-nature relationship: the case of the agrico-pastoral Iraqw of the old Mbulu District', in P.G. Forster and S. Maghimbi (eds) *The Tanzanian Peasantry: Economy in Crisis*, 45-57. Aldershot: Avebury.

—— 1999. 'May the spider web blind witches and wild animals: local knowledge and the political ecology of natural resource use in Iraqwland, Tanzania, 1900-1985'. PhD thesis, Boston University.

Leach, H.M. 1999. 'Intensification in the Pacific. A critique of the archaeological criteria and their application', *Current Anthropology* 40: 311-39.

Leach, M. and Mearns, R. (eds) 1996. *The Lie of the Land: Challenging Received Wisdom on the African Environment*. Oxford: James Currey for the IAI.

Lindblade, K.A., Carswell, G. and Tumuhairwe, J.K. 1998. 'Mitigating the relationship between population growth and land degradation: land use change and farm management in southwestern Uganda', *Ambio* 27: 565-71.

Lindström, J. 1988. 'FTP in Babati, Tanzania'. Working Paper No. 83, International Rural Development Centre. Uppsala: Swedish University of Agricultural Sciences.

Little, P.D. 1985. 'Social differentiation and pastoralist sedentarization in Northern Kenya', *Africa* 55: 356-61.

—— 1992. *The Elusive Granary: Herder, Farmer and State in Northern Kenya*. Cambridge University Press.

Liwenga, E. and Lyimo, J. 1996. 'The relevance of traditional farming systems in sustaining agricultural production: the case of the Iraqw farming system'. Institute of Resource Assessment, University of Dar es Salaam.

Loiske, V.-M. 1995. *The Village that Vanished*. Meddelanden series B 94, Department of Human Geography, Stockholm University.

—— n.d. 'Traditional natural resource management institutions among the Iraqw in Tanzania'. Environment and Development Studies Unit, School of Geography, Stockholm University.

Ludwig, H.-D. 1968. 'Permanent farming on Ukara: the impact of land shortage on husbandry practices', in H. Ruthenberg (ed.) *Smallholder Farming and Smallholder Development in Tanzania: Ten Case Studies*, 87-135. Munich: Weltforum.

Maddox, G., Giblin, J.L. and Kimambo, I.N. (eds) 1996. *Custodians of the Land. Ecology and Culture in the History of Tanzania*. London: James Currey.

Maggs, T. 1995. 'From Marateng to Marakwet: islands of agricultural intensification in Eastern and Southern Africa'. Abstracts for 10th Congress of the Pan African Association for Prehistory. Harare: University of Zimbabwe and National

Museums and Monuments of Zimbabwe.

Maher, C. 1935. 'A partial survey of the agricultural resources and potentialities of the West Suk Reserve, together with notes on the general state of preservation, agriculturally and pastorally of the area'. Typescript, 32 pp., Rhodes House Library, Oxford.

Manshard, W. 1968. *Einführung in die Agrargeographie der Tropen*. Mannheim: Bibliographisches Institut.

Masao, F.T. 1974. 'The irrigation system in Uchagga: an ethnohistorical approach', *Tanzania Notes and Records* 75: 1-8.

Massam, J.A. 1927. *The Cliff Dwellers of Kenya*. London: Seeley, Service.

Mbulu District Book (various notes by British district officials, 1916-61). Rhodes House Library, Oxford (microfilm).

McCann, J.C. 1995. *People of the Plow: an Agricultural History of Ethiopia, 1800–1990*. Madison: University of Wisconsin Press.

—— 1999. *Green Land, Brown Land, Black Land: an Environmental History of Africa, 1800–1990*. Portsmouth, NH and Oxford: Heinemann and James Currey.

McCown, R.L., Haaland, G. and Haan, C. de. 1979. 'The interaction between cultivation and livestock in semi-arid Africa', in A.E. Hall, G.H. Cannell and H.W. Lawton (eds) *Agriculture in Semi-Arid Environments*, 297-332. Berlin, Heidelberg, New York: Springer.

Meek, C.I. 1953. 'Stock reduction in the Mbulu Highlands, Tanganyika', *Journal of African Administration* 5: 158-66.

Meindertsma, J.D. and Kessler, J.J. (eds) 1997. 'Towards better use of environmental resources. A planning document of Mbulu and Karatu Districts, Tanzania'. Mbulu District Council.

Messeret Lejebo 1990. 'A study of the technological, social, socio-economic and environmental constraints affecting the performance of Ethiopian agriculture: with special reference to drought and famine crises in Konso'. *Giessener Schriften für Wirtschafts- und Regionalsociologie*, Heft 18.

Ministry of Agriculture, People's Democratic Republic of Ethiopia and Tahal Consulting Engineers Ltd (in Association with Shawel Consul International) 1988. 'Study of traditional conservation practices'.

Moore, H. 1986. *Space, Text and Gender: an Anthropological Study of the Marakwet of Kenya*. Cambridge University Press.

Moore, S.F. 1986. *Social Facts and Fabrications: 'Customary' Law on Kilimanjaro, 1880–1980*. Cambridge University Press.

Morgan, W.T.W. 1988. 'Tamilnad and eastern Tanzania: comparative regional geography and the historical development process', *Geographical Journal* 154: 69-86.

Morrison, K.D. 1994. 'The intensification of production: archaeological approaches', *Journal of Archaeological Method and Theory* 1: 111-59.

—— 1996. 'Typological schemes and agricultural change: beyond Boserup in precolonial South India', *Current Anthropology* 37: 583-608.

Mortimore, M.J. 1989. *Adapting to Drought: Farmers, Famines and Desertification in West Africa*. Cambridge University Press.

—— 1998. *Roots in the African Dust: Sustaining the Sub-Saharan Drylands*. Cambridge University Press.

—— 2002. 'Development and change in Sahelian dryland agriculture', in I.

Livingstone and D. Belshaw (eds) *Renewing Development in sub-Saharan Africa: Policy, Performance and Prospects*, 135-52. London: Routledge.

Mortimore, M.J. and Adams, W.M. 1999. *Working the Sahel: Environment and Society in Northeast Nigeria*. London: Routledge.

Mortimore, M.J. and Tiffen, M. 1995. 'Population and environment in time perspective: the Machakos story', in T. Binns (ed.) *People and Environment in Africa,* 69–89. Chichester: Wiley.

Mung'ong'o, C. 1995. *Social Processes and Ecology in the Kondoa Irangi Hills, Central Tanzania.* Department of Human Geography, Stockholm University.

—— 1999. 'Sustainable agriculture in semi-arid Tanzania: some sociological insights from Kondoa-Irangi hills, Ukara island, Matengo plateau and Mbulu highlands', in J. Boesen, I.S. Kikula and F.P. Maganga (eds) *Sustainable Agriculture in Semi-arid Tanzania,* 40-64. Dar es Salaam University Press.

Murdock, G.P. 1959. *Africa: Its Peoples and their Culture History.* New York: McGraw-Hill.

Muriuki, G. 1974. *A History of the Kikuyu* 1500–1900. Nairobi: Oxford University Press.

Murton, J. 1999. 'Population growth and poverty in Machakos District, Kenya', *Geographical Journal* 165: 37-46.

Nandwa, S.M. and Bekunda, M.A. 1998. 'Research on nutrient flows and balances in East and Southern Africa: state-of-the-art', *Agriculture, Ecosystems and Environment* 71: 5-18.

National Soil Service (NSS). 1994. 'Land resources inventory and land suitability assessment of Mbulu District, Arusha Region, Tanzania'. NSS, Tanga, Tanzania, Reconnaissance Soil Survey, Report No. R5, 7 volumes, 8 separate maps.

Netting, R.McC. 1993. *Smallholders, Householders: Farm Families and the Ecology of Intensive Sustainable Agriculture.* Stanford, CA: Stanford University Press.

Niemeijer, D. 1996. 'The dynamics of African agricultural history: Is it time for a new development paradigm?', *Development and Change* 27: 87-110.

Nowack, E. 1942. 'Die Bevölkerungsverteilung in Deutsch-Ostafrika und ihre Uhrsachen', *Petermanns Geographische Mitteilungen*, 88: 367-9.

Nowrojee, B., Manby, B. and An-Na'im, A. 1993. *Divide and Rule: State-sponsored Ethnic Violence in Kenya.* New York: Human Rights Watch.

Nurse, D. and Rottland, F. 1993. 'The history of Sonjo and Engaruka: a linguists' view', *Azania* 28: 1-5.

Nyberg, H. 1999. *Tenure Change – Driving Forces and Local Response: Examples from North-Western Kenya.* Department of Social and Economic Geography, Lund University.

Padoch, C. 1985. 'Labour efficiency and intensity of land use in rice production: an example from Kalimantan', *Human Ecology* 13: 271-89.

Peristiany, J.G. 1951. 'The age-set system of the pastoral Pokot', *Africa* 21: 188-206, 279-302.

Pike, A.G. 1965. 'Kilimanjaro and the furrow system', *Tanganyika Notes and Records* 64: 95-6.

Pingali, P., Bigot, Y. and Binswanger, H. 1987. *Agricultural Mechanisation and the Evolution of Farming Systems in Sub-Saharan Africa.* Baltimore: Johns Hopkins University Press.

Price, D.H. 1994. 'Wittfogel's neglected hydraulic/hydroagricultural distinction', *Journal of Anthropological Research* 50: 187-204.

Raikes, P.L. 1975. 'The development of mechanized commercial wheat production in North Iraqw, Tanzania'. PhD thesis, Stanford University.

Ranger, T. 1983. 'The invention of tradition in colonial Africa', in E. Hobsbawm and T. Ranger (eds) *The Invention of Tradition,* 211-62. Cambridge University Press.

—— 1993. 'The invention of tradition revisited: the case of colonial Africa', in T. Ranger and O. Vaughan (eds) *Legitimacy and the State in Twentieth-Century Africa,* 62-111. London: Macmillan.

Raynaut, C. (with E. Grégoire, P. Janin, J. Koechlin and P.L. Delville) 1998. *Societies and Nature in the Sahel.* London: Routledge.

Reid, A. 1998. 'The antiquity of the ethnic divide in Great Lakes Africa'. Paper for the 4th World Archaeological Congress, Cape Town, 1999.

Reid, A. and Young, R. 2000. 'Pottery abrasion and the preparation of African grains', *Antiquity* 74: 101-11.

Rekdal, O, B. 1999. *The invention by tradition: Creativity and change among the Iraqw of northern Tanzania.* Dr. Polit. thesis in Social Anthropology, University of Bergen.

Richards, P. 1985. *Indigenous Agricultural Revolution: Ecology and Food Production in West Africa.* London: Hutchinson.

—— 1986. *Coping with Hunger: Hazard and Experiment in an African Rice Farming System.* London: Allen and Unwin.

—— 1989. 'Agriculture as performance', in R. Chambers, A. Pacey and L.A. Thrupp (eds) *Farmer First: Farmer Innovation and Agricultural Research*, 39-43. London: Intermediate Technology Publications.

Rigby, P. 1985. *Persistent Pastoralists.* London: Zed Books.

Rocheleau, D., Steinberg, P.E. and Bengamin, P.A. 1995. 'Environment, development, crisis and crusade: Ukambani, Kenya, 1980–1990', *World Development* 23: 1037-51.

Rodman, M. 1992. 'Empowering place: multilocality and multivocality', *American Anthropologist* 94: 640-56.

Roe, E. 1991. 'Development narratives, or making the best of blueprint development', *World Development* 19: 287-300.

Rottland, F. 1983. 'Marakwet dialects: synchronic and diachronic aspects', in B.E. Kipkorir, R.C. Soper and J.W. Ssennyonga (eds) *Kerio Valley: Past, Present and Future,* 139-46. University of Nairobi, Institute of African Studies.

Rowe, E.G. 1947. 'Notes on Mbulu District, Northern Province, Tanganyika Territory (for Lord Hailey)', in Hailey Papers: Notes on Masai, Mbulu and Moshi Districts. Unpublished collection of notes, Public Records Office, London, CO 1018/71.

Ruthenberg, H. 1971. *Farming Systems in the Tropics.* Oxford: Clarendon Press.

Sahlins, M. 1963. 'Poor Man, Rich Man, Big-man, Chief: Political types in Melanesia and Polynesia', *Comparative Studies in Society and History* 5: 285-303.

—— 1974. *Stone Age Economics.* London: Tavistock.

Schlippe, P. de. 1956. *Shifting Cultivation in Africa: the Zande System of Agriculture.* London: Routledge and Kegan Paul.

Schneider, H.K. 1979. *Livestock and Equality in East Africa.* Bloomington: Indiana University Press.

Schoenbrun, D.L. 1999. *A Green Place, a Good Place: Agrarian Change, Gender and Social Identity in the Great Lakes Region to the 15th century.* Oxford and Portsmouth, NH: James Currey, Heinemann.

Schultz, J. 1971. *Agrarlandschaftliche Veränderungen in Tanzania.* Munich: Weltforum.

Scoones, I. 1995. *Hazards and Opportunities: Farming Livelihoods in Dryland Africa: Lessons from Zimbabwe.* London: Zed Books.

—— 1997a. 'The dynamics of soil fertility change: historical perspectives on environmental transformations from Zimbabwe', *Geographical Journal* 163: 161-9.

—— 1997b. 'Landscapes, fields and soils: understanding the history of soil fertility management in southern Zimbabwe', *Journal of Southern African Studies* 23: 615-34.

Scoones, I. and Toulmin, C. 1998. 'Soil nutrient balances: what use for policy?', *Agriculture, Ecosystems and Environment* 7: 255-67.

Scoones, I., Reij C. and Toulmin, C. 1996. 'Indigenous soil and water conservation in Africa', in C. Reij., I. Scoones and C. Toulmin (eds) *Sustaining the Soil: Traditional Soil and Water Conservation in Africa,* 1-27. London: Earthscan.

Scott, J. 1976. *The Moral Economy of the Peasant: Rebellion and Subsistence in Southeast Asia.* New Haven, CT: Yale University Press.

Service, E.R. 1975. *Origins of the State and Civilization.* New York: Norton.

Shako Otto. 1993. 'The Nine Clans of Konso', *Sociology Ethnology Bulletin* 1: 80-92.

Sheridan, M.J. 2001. 'Cooling the land: the political ecology of the North Pare mountains, Tanzania'. PhD dissertation. Boston University.

—— 2002. 'An irrigation intake is like a uterus: culture and agriculture in precolonial North Pare, Tanzania', *American Anthropologist* 104: 79-92.

Shipton, P.M. 1984a. 'Strips and patches: a demographic dimension in some African land-holding and political systems', *Man* 19: 613-34.

—— 1984b. 'Lineage and locality as antithetical principles in East African systems of land tenure', *Ethnology* 23: 117-32.

Snyder, K.A. 1993. '"Like water and honey": moral ideology and the construction of community among the Iraqw of northern Tanzania'. PhD thesis, Yale University.

—— 1996. 'Agrarian change and land-use strategies among Iraqw farmers in northern Tanzania', *Human Ecology* 24: 315-40.

—— 1997. 'Elders' authority and women's protest: the *masay* ritual and social change among the Iraqw of Tanzania', *Journal of the Royal Anthropological Institute* NS 3: 561-76.

—— 1999. 'Gender ideology, and the domestic and public domains among the Iraqw', in H.L. Moore, T. Sanders, and B. Kaare (eds) *Those Who Play with Fire: Gender, Fertility and Transformation in East and Southern Africa,* 225-53. London: Athlone.

Soper, R. 1983. 'A survey of the irrigation systems of the Marakwet', in B.E. Kipkorir, R.C. Soper and J.W. Ssennyonga (eds) *Kerio Valley: Past, Present and Future,* 75-95. University of Nairobi, Institute of African Studies.

—— 1996. 'The Nyanga terrace complex of eastern Zimbabwe', *Azania* 31: 1-36

—— 2000. 'The agricultural landscape of the Nyanga area, Zimbabwe', in G. Barker and D. Gilbertson (eds) *The Archaeology of Drylands: Living at the Margin,* 201-19. London: Routledge.

—— 2002. *Nyanga: Ancient Fields, Settlements and Agricultural History in Zimbabwe*. London: British Institute in Eastern Africa, Memoir 16.

Southall, A.W. 1961. 'Population movements in East Africa', in K.M. Barbour and R.M. Prothero (eds) *Essays on African Population,* 157-92. London: Routledge & Kegan Paul.

—— 1970. 'The illusion of tribe', in P.C.W. Gutkind (ed.) *The Passing of Tribal Man in Africa*, 28-50. Leiden: Brill.

Spear, T. 1993. 'Introduction' and 'Being "Maasai", but not "people of cattle": Arusha agricultural Maasai in the nineteenth century', in T. Spear and R. Waller (eds) *Being Maasai: Ethnicity and Identity in East Africa*, 1-18, 120-36. London: James Currey.

—— 1997. *Mountain Farmers: Moral Economies of Land and Agricultural Development in Arusha and Meru*. Oxford: James Currey.

Spencer, P. 1998. *The Pastoral Continuum: the Marginalization of Tradition in East Africa*. Oxford: Clarendon.

Ssennyonga, J.W. 1983. 'The Marakwet irrigation system as a model of a systems-approach to water management', in B.E. Kipkorir, R.C. Soper and J.W. Ssennyonga (eds) *Kerio Valley: Past, Present and Future,* 96-111. University of Nairobi, Institute of African Studies.

Stone, G.D. 1996. *Settlement Ecology: the Social and Spatial Organization of Kofyar Agriculture*. Tucson: University of Arizona Press.

Straube, H. 1963. *Westkuschitische Völker Süd-Äthiopiens*. Stuttgart: Kohlhammer for Frobenius-Institut (Goethe-Universität, Frankfurt-am-Main).

Sturdy, D. 1936. 'Agricultural notes', *Tanganyika Notes and Records* 1: 52-4.

Summers, R. 1958. *Inyanga: Prehistoric Settlements in Southern Rhodesia*. Cambridge University Press.

Sutton, J.E.G. 1973. *The Archaeology of the Western Highlands of Kenya*, Memoir No. 3. Nairobi: British Institute in Eastern Africa.

—— 1978. 'Engaruka and its waters', *Azania* 13:37-70.

—— 1984. 'Irrigation and soil conservation in African agricultural history: with a reconsideration of the Inyanga terracing (Zimbabwe) and Engaruka irrigation works (Tanzania)', *Journal of African History* 25: 25-41.

—— 1985. 'Irrigation and terracing in Africa: intensification, specialisation or overspecialisation?' in I.S. Farrington (ed.) *Prehistoric Intensive Agriculture in the Tropics*. Oxford: BAR International series 232.

—— 1986. 'The irrigation and manuring of the Engaruka field system', *Azania* 21: 27-51.

—— 1987. 'Hyrax Hill and the Sirikwa', *Azania* 22: 1-36.

—— 1989. 'Towards a history of cultivating the fields', *Azania* 24: 99-112.

—— 1990. *A Thousand Years of East Africa*. Nairobi: British Institute in Eastern Africa.

—— 1993. 'Becoming Maasailand', in T. Spear and R. Waller (eds) *Being Maasai: Ethnicity and Identity in East Africa,* 38-60. London: James Currey.

—— 1994. 'The Sirikwa and the Okiek in the History of the Kenya Highlands', *Kenya Past and Present* 26: 35-40.

—— 1998. 'Engaruka: an irrigation agricultural community in northern Tanzania before the Maasai', *Azania* 33: 1-37.

Tagseth, M. 2000. 'Knowledge and development in *mifongo* irrigation systems. Three case studies from Mt Kilimanjaro, Tanzania'. MA thesis, Norwegian University

of Science and Technology.
Tengö, M. 1999. 'Integrated nutrient management and farming practices in the agro-ecological system of Mama Issara'. Unpublished report, Department of Systems Ecology, Stockholm University.
Thomas, G.L. 1977. 'Baseline information and situational overview requisite to the design of integrated rural development projects in Mbulu District, Tanzania'. Unpublished Report for USAID, Tanzania.
Thomson, J. 1887. *Through Masai Land*, 2nd edn (Cass reprint 1968). London: Low, Marston, Searle and Rivington.
Thornton, R.J. 1980. *Space, Time, and Culture among the Iraqw of Tanzania*. New York: Academic Press.
Tiffen, M. and Mortimore, M.J. 1992. 'Environment, population growth and production in Kenya: a case study of Machakos District', *Development Policy Review* 10: 359-87.
—— 1994. 'Malthus controverted: the role of capital and technology in growth and environmental recovery in Kenya', *World Development* 22: 997-1010.
Tiffen, M., Mortimore, M.J. and Gichuki, F. 1994. *More People, Less Erosion: Environmental Recovery in Kenya*. Chichester: Wiley.
Trewartha, G. and Zelinsky, W. 1954. 'Population patterns in tropical Africa', *Annals of the Association of American Geographers* 44: 136-62.
Turner, B.L., Hyden, G. and Kates, R.W. 1993. *Population Growth and Agricultural Change in Africa*. Gainesville: University Press of Florida.
Waller, R. 1985. 'Ecology, migration, and expansion in East Africa', *African Affairs* 84: 347-70.
—— 1988. 'Emutai: crisis and response in Maasailand 1883–1902', in D.H. Johnson and D.M. Anderson (eds) *The Ecology of Survival: Case Studies from Northeast African History*, 73-112. London and Boulder, CO: Crook.
Watson, E.E. 1997. 'Ritual leaders and agricultural resources in south-western Ethiopia: *Poqallas*, land and labour in Konso', in K. Fukui, E. Kurimoto and M. Shigeta (eds) *Ethiopia in Broader Perspective: Papers of the XIIIth International Conference of Ethiopian Studies in Kyoto, Japan*, 652-69. Kyoto: Shakodo Booksellers.
—— 1998. 'Ground truths: land and power in Konso, Ethiopia'. PhD thesis, Cambridge University.
—— 2002. 'Capturing a local elite: the Konso honeymoon', in W. James, D. Donham, A. Triulzi and E. Kurimoto (eds) *Remapping Ethiopia: Socialism and After.* Oxford: James Currey.
Watson, E.E., Adams, W.M. and Mutiso, S.K. 1998. 'Indigenous irrigation, agriculture and development, Marakwet, Kenya', *Geographical Journal* 64: 67-84.
Watson, E.E. and Lakew Regassa 2001. 'Konso', in D. Freeman and A. Pankhurst (eds) *Living on the Edge: Marginalised Minorities of Craftworkers and Hunters in Southern Ethiopia*, 246-64. Addis Ababa: Department of Sociology and Social Adminstration.
Werther, C.W. 1898. *Die mittleren Hochlander des nördlichen Deutsch-Ostafrika*. Berlin: Paetel.
Westphal, E. 1975. *Agricultural Systems in Ethiopia*. Wageningen: Agricultural University.
Widgren, M. 2000. 'Islands of intensive agriculture in African drylands: towards an explanatory framework', in G. Barker and D. Gilbertson (eds) *The Archaeology of Drylands: Living at the Margin*, 252-67. London: Routledge.

Wiggins, S. 1995. 'Changes in African farming systems between the mid-1970s and the mid-1980s', *Journal of International Development* 7: 807-48.

Winter, E.H. and Molyneaux, L. 1963. 'Population patterns and problems among the Iraqw', *Ethnology* 2: 490-505.

Wittfogel, K. 1957. *Oriental Despotism.* New Haven, CT: Yale University Press.

Yanda, P. and Mohamed, S.A. 1990. 'The environmental impact of irrigation schemes: the case of Mto wa Mbu, Arusha, Tanzania: a reconnaissance survey'. Dar es Salaam: Institute of Resource Assessment.

Index